DYNAMICS OF HUMAN GAIT

Christopher L. Vaughan, PhD
University of Virginia

Brian L. Davis, PhD
Cleveland Clinic Foundation

Jeremy C. O'Connor, BSc
University of the Western Cape

Human Kinetics Publishers
Champaign, Illinois

Library of Congress Cataloging-in-Publication Data

Vaughan, Christopher L.
  Dynamics of human gait / Christopher L. Vaughan, Brian L. Davis, Jeremy C. O'Connor.
    p.   cm.
  Includes bibliographical references and index.
  ISBN 0-87322-368-3
  1. Gait in humans.   I. Davis, Brian L., 1960-   II. O'Connor, Jeremy C., 1968- .   III. Title.
  [DNLM: 1. Gait--physiology.   WE 103 V365d]
  QP310.W3V38   1992
  612.7'6--dc20
  DNLM/DLC
  for Library of Congress   92-1471
                                                                   CIP

ISBN: 0-87322-368-3
ISBN: 0-87322-370-5   3 1/2'' disk pkg
ISBN: 0-87322-371-3   5 1/4'' disk pkg

Copyright © 1992 by Christopher L. Vaughan

All rights reserved. Except for use in a review, the reproduction or utilization of this work in any form or by any electronic, mechanical, or other means, now known or hereafter invented, including xerography, photocopying, and recording, and in any information storage and retrieval system, is forbidden without the written permission of the publisher.

The terms IBM, PC, XT, AT, and PS/2 are all trademarks of International Business Machines Corporation of Armonk, NY.

Acquisitions Editor: Rick Frey, PhD
Developmental Editors: Marie Roy and Larret Galasyn-Wright
Assistant Editors: Elizabeth Bridgett, Valerie Hall, and Julie Swadener
Copyeditor: Dianna Matlosz
Proofreader: Christine Drews
Indexer: Barbara Cohen
Production Director: Ernie Noa
Typesetter: Sandra Meier
Text Design: Keith Blomberg
Text Layout: Tara Welsch
Cover Design: Jack Davis
Illustrations: Ron Ervin and Gretchen Walters
Printer: Versa Press

Printed in the United States of America

10 9 8 7 6 5 4 3 2 1

Human Kinetics Publishers
Box 5076, Champaign, IL 61825-5076
1-800-747-4457

Canada Office:
Human Kinetics Publishers, Inc.
P.O. Box 2503, Windsor, ON N8Y 4S2
1-800-465-7301 (in Canada only)

Europe Office:
Human Kinetics Publishers (Europe) Ltd.
P.O. Box IW14
Leeds LS16 6TR
England
0532-781708

This book is dedicated to our families:

Joan, Bronwyn, and Gareth Vaughan;
Tracy and Sean Davis;
and
the O'Connor Family.

Contents

About *Gait Analysis Laboratory*   vii
About *Dynamics of Human Gait*   ix
Acknowledgments   xi

Chapter 1   In Search of the Homunculus   1
Top-Down Analysis of Gait   2
Measurements and the Inverse Approach   4
Summary   6

Chapter 2   The Three-Dimensional and Cyclic Nature of Gait   7
Periodicity of Gait   8
Parameters of Gait   12
Summary   14

Chapter 3   Integration of Anthropometry, Displacements, and Ground Reaction Forces   15
Body Segment Parameters   16
Linear Kinematics   22
Centers of Gravity   29
Angular Kinematics   32
Dynamics of Joints   35
Summary   43

Chapter 4   Muscle Actions Revealed Through Electromyography   45
Back to Basics   45
Phasic Behavior of Muscles   52
Relationship Between Different Muscles   55
Summary   61

Chapter 5 Clinical Gait Analysis—A Case Study — 63
Experimental Methods — 64
Results and Discussion — 65
Summary — 76

Appendix A Dynamic Animation Sequences — 77

Appendix B Detailed Mathematics Used in Gaitmath — 83

Appendix C Commercial Equipment for Gait Analysis — 107

References — 127
Index — 133

About *Gait Analysis Laboratory*

Gait Analysis Laboratory has its origins in the Department of Biomedical Engineering of Groote Schuur Hospital and the University of Cape Town. It was in the early 1980s that the three of us first met to collaborate on the study of human walking. Our initial efforts were simple and crude. Our two-dimensional analysis of children with cerebral palsy and nondisabled adults was performed with a movie camera, followed by tedious manual digitizing of film in an awkward minicomputer environment. We concluded that others traveling this road should have access—on a personal computer—to material that conveys the essential three-dimensional and dynamic nature of human gait. This package is a result of that early thinking and research.

There are three parts to *Gait Analysis Laboratory*: this book, *Dynamics of Human Gait*, and the *GaitLab* software and manual. In the book we establish a framework for gait analysis and explain our theories and techniques. One of the notable features is the detailed animation sequence that begins in Appendix A. These walking figures are analog counterparts to the digital animation presented in *GaitLab*, the IBM-compatible software that is the heart of this package. *GaitLab*'s sizable data base lets you explore and plot more than 250 combinations of the basic parameters used in gait analysis. These can be displayed in a variety of combinations, both graphically and (to a lesser degree) with stick figure animation. An added feature of *GaitLab* is Animate, a single cycle, closed-loop animation sequence of a walking man. Finally, to help you get our programs up and running, we have written the *GaitLab* software manual.

We've prepared this package with the needs of all students of human movement in mind. Our primary objective has been to make the theory and tools of 3-D gait analysis available to the person with a basic knowledge of mechanics and anatomy and access to a personal computer. In this way we believe that this package will appeal to a wide audience. In particular, the material should be of interest to the following groups:

- Students and teachers in physical education and physical therapy
- Clinicians in orthopaedic surgery, physical therapy, podiatry, rehabilitation, neurology, and sports medicine
- Researchers in biomechanics, kinesiology, biomedical engineering, and the movement sciences in general

Whatever your specific area of interest, after working with *Gait Analysis Laboratory* you should have a much greater appreciation for the human locomotor apparatus, particularly how we all manage to coordinate movement in three dimensions. These powerful yet affordable tools were designed to provide new levels of access to the complex data generated by a modern gait analysis laboratory. By making this technology available we hope to deepen your understanding of the dynamics of human gait.

About *Dynamics of Human Gait*

This book was created as a companion to *GaitLab*. Our intent was to introduce gait analysis, not provide a comprehensive guide. We try to serve readers with diverse experience and areas of interest by discussing the basics of human gait as well as some of the theoretical, biomechanical, and clinical aspects.

In chapter 1 we take you in search of the homunculus, the little being inside each of us who makes our walking patterns unique. We represent the walking human as a series of interconnected systems—neural, muscular, skeletal, mechanical, and anthropometric—that form the framework for detailed gait analysis.

The three-dimensional and cyclical nature of human gait is described in chapter 2. We also explain how many of the relevant parameters can be expressed as a function of the gait cycle, including kinematics (e.g., height of lateral malleolus), kinetics (e.g., vertical ground reaction force), and muscle activity (e.g., EMG of rectus femoris).

In chapter 3 we show you how to use the framework constructed in the first two chapters to integrate anthropometric, 3-D kinematic, and 3-D force plate data. For most readers this will be an important chapter—it is here that we suggest many of the conventions we believe to be lacking in three-dimensional gait analysis. Although conceptually rigorous, the mathematical details are kept to a minimum to make the material accessible to all students of human motion. (For the purists interested in these details, that information is in Appendix B.)

In chapter 4 we describe the basics of electromyography (EMG) and how it reveals the actions of the various muscle groups. We discuss some of the techniques involved and then illustrate the phasic behavior of muscles during the gait cycle and describe how these signals may be statistically analyzed.

One of the purposes of this book is to help clinicians assess the gaits of their patients. Chapter 5 presents a case study of a 23-year-old man with cerebral palsy. We have a complete set of 3-D data for him that can be processed and analyzed in *GaitLab*.

Beginning in Appendix A we use illustrated animation sequences to emphasize the dynamic nature of human gait. By carefully fanning the pages of the appendixes, you can get a feel for the way the human body integrates muscle activity, joint moments, and ground reaction forces to produce a repeatable gait pattern. These sequences bring the walking subject to life and may provide you with new insights.

The detailed mathematics used to integrate anthropometry, kinematics, and force plate data and to generate 3-D segment orientations, and 3-D joint forces and moments are presented in Appendix B. This material, which is the basis for the Gaitmath program used in *GaitLab*, has been included for the sake of completeness. It is intended for researchers who may choose to include some of the equations and procedures in their own work.

The various pieces of commercially available equipment that may be used in gait analysis are described and compared (including approximate prices) in Appendix C.

Dynamics of Human Gait provides a solid foundation for those new to gait analysis, while at the same time addressing advanced mathematical techniques used for computer modeling and clinical study. As the first part of *Gait Analysis Laboratory*, the book should act as a primer for your exploration within the *GaitLab* environment. We trust you will find the material both innovative and informative.

Acknowledgments

We are grateful to all those who have enabled us to add some diversity to our book. It is a pleasure to acknowledge the assistance of Dr. Peter Cavanagh, director of the Center for Locomotion Studies (CELOS) at Pennsylvania State University, who provided the plantar pressure data used for our animation sequence, and Mr. Ron Ervin, who drew the human figures used in the sequence.

Dr. Andreas von Recum, professor and head of the Department of Bioengineering at Clemson University, and Dr. Michael Sussman, chief of Pediatric Orthopaedics at the University of Virginia, provided facilities, financial support, and substantial encouragement during the writing of the text.

The three reviewers, Dr. Murali Kadaba of Helen Hayes Hospital, Dr. Stephen Messier of Wake Forest University, and Dr. Cheryl Riegger of the University of North Carolina, gave us substantial feedback. Their many suggestions and their hard work and insights have helped to make this a better book.

We are especially grateful to Mrs. Nancy Looney and Mrs. Lori White, who helped with the early preparation of the manuscript.

Appendix C, "Commercial Equipment for Gait Analysis," could not have been undertaken without the interest and cooperation of the companies mentioned.

The major thrust of *Gait Analysis Laboratory*, the development of *GaitLab*, took place in June and July of 1988 in Cape Town. We especially thank Dr. George Jaros, professor and head of the Depatment of Biomedical Engineering at the University of Cape Town and Groote Schuur Hospital. He established an environment where creativity and collaboration flourished. We also acknowledge the financial support provided by the university, the hospital, and the South African Medical Research Council.

Much of the conceptual framework for *Gait Analysis Laboratory* was developed during 1983-84 in England at the University of Oxford's Orthopaedic Engineering Centre (OOEC). Dr. Michael Whittle, deputy director, and Dr. Ros Jefferson, mathematician, provided insight and encouragement during this time. They have maintained an interest in our work and recently shared some of their kinematic and force plate data, which are included in *GaitLab*.

The data in chapters 3 and 5 were provided by Dr. Steven Stanhope, director, and Mr. Tom Kepple, research scientist, of the Biomechanics Laboratory at the National Institutes of Health in Bethesda, MD; and by Mr. George Gorton, technical director, and Ms. Patty Payne, research physical therapist, of the Motion Analysis Laboratory at the Children's Hospital in Richmond, VA. Valuable assistance was rendered by Mr. Francisco Sepulveda, graduate student in bioengineering, in the gathering and analysis of the clinical data.

Finally, it is a pleasure to acknowledge the efforts of the staff at Human Kinetics. We make special mention of Dr. Rainer Martens, publisher, Dr. Rick Frey, director of HK Academic Book Division, and Ms. Marie Roy and Mr. Larret Galasyn-Wright, developmental editors, who have been enthusiastic, supportive, and above all, patient.

CHAPTER 1

In Search of the Homunculus

> *Homunculus: An exceedingly minute body that according to medical scientists of the 16th and 17th centuries, was contained in a sex cell and whose preformed structure formed the basis for the human body.*
>
> Stedman's Medical Dictionary

When we think about the way in which the human body walks, the analogy of a marionette springs to mind. Perhaps the puppeteer who pulls the strings and controls our movements is a homunculus, a supreme commander of our locomotor program. Figure 1.1, reprinted from Inman, Ralston, and Todd (1981), illustrates this point in a rather humorous but revealing way. Though it seems simplistic, we can build on this idea and create a structural framework or model that will help us to understand the way gait analysis should be performed.

Figure 1.1 A homunculus controls the dorsiflexors and plantar flexors of the ankle, and thus determines the pathway of the knee. *Note.* From *Human Walking* (p. 11) by V.T. Inman, H.J. Ralston, and F. Todd, 1981, Baltimore: Williams & Wilkins. Copyright 1981 by Williams & Wilkins. Reprinted by permission.

Top–Down Analysis of Gait

Dynamics of Human Gait takes a *top–down* approach to the description of human gait. The process that we are most interested in starts as a nerve impulse in the central nervous system and ends with the generation of ground reaction forces. The key feature of this approach is that it is based on *cause* and *effect*.

Sequence of Gait-Related Processes

We need to recognize that locomotor programming occurs in supraspinal centers and involves the conversion of an *idea* into the pattern of muscle activity that is necessary for walking (Enoka, 1988). The neural output that results from this supraspinal programming may be thought of as a central locomotor command being transmitted to the brainstem and spinal cord. The execution of this command involves two components:

1. Activation of the lower neural centers, which subsequently establish the sequence of muscle activation patterns
2. Sensory feedback from muscles, joints, and other receptors that modifies the movements.

This interaction between the central nervous system, peripheral nervous system, and musculoskeletal effector system is illustrated in Figure 1.2 (Jacobsen & Webster, 1977). (For the sake of clarity, the feedback loops have not been included in this figure.) The muscles, when activated, develop tension, which in turn generates forces at, and moments across, the synovial joints.

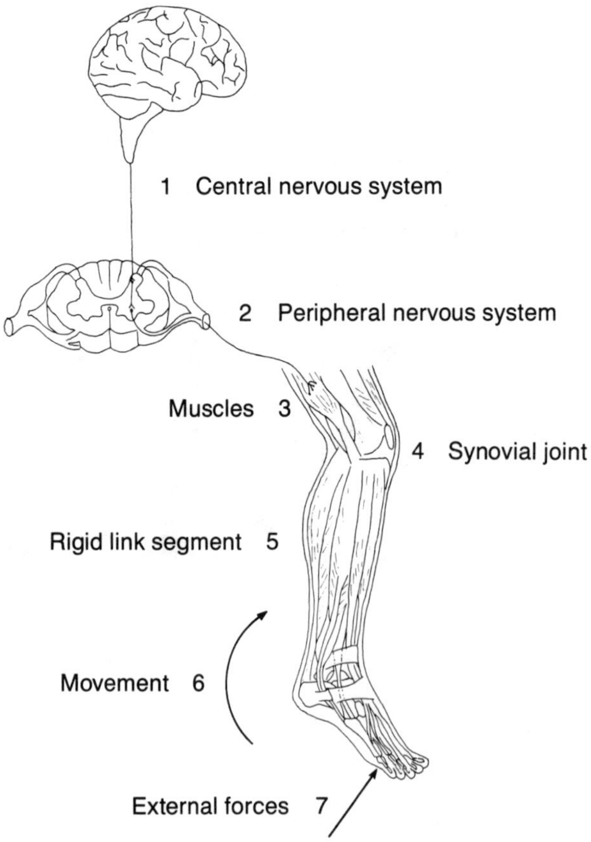

Figure 1.2 The seven components that form the functional basis for the way in which we walk. This top-down approach constitutes a cause-and-effect model.

1 Central nervous system
2 Peripheral nervous system
3 Muscles
4 Synovial joint
5 Rigid link segment
6 Movement
7 External forces

The joint forces and moments cause the rigid skeletal links (segments such as the thigh, calf, foot, etc.) to move and to exert forces on the external environment.

The sequence of events that must take place for walking to occur may be summarized as follows:

1. Registration and activation of the gait command in the central nervous system
2. Transmission of the gait signals to the peripheral nervous system
3. Contraction of muscles that develop tension
4. Generation of forces at, and moments across, synovial joints
5. Regulation of the joint forces and moments by the rigid skeletal segments based on their anthropometry
6. Displacement (i.e., movement) of the segments in a manner that is recognized as functional gait
7. Generation of ground reaction forces

These seven links in the chain of events that result in the pattern of movement we readily recognize as human walking are illustrated in Figure 1.3.

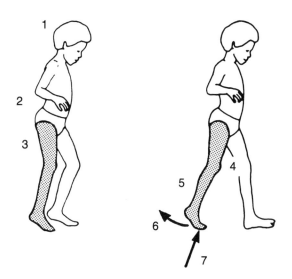

Figure 1.3 The sequence of seven events that lead to walking. *Note.* This illustration of a hemiplegic cerebral palsy child has been adapted from *Gait Disorders in Childhood and Adolescence* (p. 130) by D.H. Sutherland, 1984, Baltimore: Williams & Wilkins. Copyright 1984 by Williams & Wilkins. Adapted by permission.

Clinical Usefulness of the Top–Down Approach

The model may also be used to help us

- understand pathology,
- determine methods of treatment, and
- decide on which methods we should use to study a patient's gait.

For example, a patient's lesion could be at the level of the central nervous system (as in cerebral palsy), in the peripheral nervous system (as in Charcot-Marie-Tooth disease), at the muscular level (as in muscular dystrophy), or in the synovial joint (as in rheumatoid arthritis). The higher the lesion, the more profound the impact on all the components lower down in the movement chain. Depending on the indications, treatment could be applied at any of the different levels. In the case of a "high" lesion, such as cerebral palsy, this could mean rhizotomy at the central nervous system level, neurectomy at the peripheral nervous system level, tenotomy at the muscular level, or osteotomy at the joint level. In assessing this patient's gait, we may choose to study the muscular activity, the anthropometry of the rigid link segments, the movements of the segments, and the ground reaction forces.

Measurements and the Inverse Approach

Measurements should be taken as high up in the movement chain as possible, so that the gait analyst approaches the *causes* of the walking pattern, not just the *effects*. As pointed out by Vaughan, Hay, and Andrews (1982), there are essentially two types of problems in rigid body dynamics. The first is the Direct Dynamics Problem in which the forces being applied (by the homunculus) to a mechanical system are known and the objective is to determine the motion that results. The second is the Inverse Dynamics Problem in which the motion of the mechanical system is defined in precise detail and the objective is to determine the forces causing that motion. This is the approach that the gait analyst pursues. Perhaps it is now clear why the title of this first chapter is "In Search of the Homunculus"!

The direct measurement of the forces and moments transmitted by human joints, the tension in muscle groups, and the activation of the peripheral and central nervous systems is fraught with methodological problems. That is why we in gait analysis have adopted the indirect or inverse approach. This approach is illustrated verbally in Figure 1.4 and mathematically in Figure 1.5. Note that four of the components in the movement chain—3, electromyography; 5, anthropometry; 6, displacement of segments; and 7, ground reaction forces—may be readily measured by the gait analyst; these have been highlighted by slightly thicker outlines in Figure 1.4. Strictly speaking, electromyography does not measure the tension in muscles, but it can give us insight into muscle activation patterns. As seen in Figure 1.5, segment anthropometry A_s may be used to generate the segment masses m_s, whereas segment displacements $\mathbf{p}_s$ may be double differentiated to yield accelerations $\mathbf{a}_s$. Ground reaction forces $\mathbf{F}_G$ are used with the segment masses and accelerations in the equations of motion which are solved in turn to give resultant joint forces and moments $\mathbf{F}_J$.

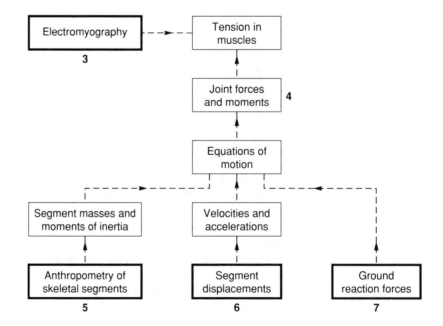

Figure 1.4 The inverse approach in rigid body dynamics expressed in words.

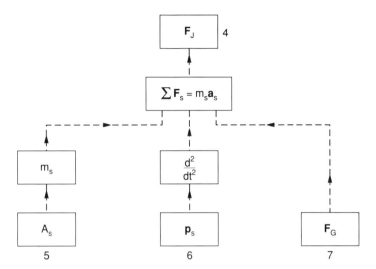

Figure 1.5 The inverse approach in rigid body dynamics expressed in mathematical symbols.

Many gait laboratories and analysts measure one or two of these components. Some measure all four. However, as seen in Figures 1.2 to 1.5, the key to understanding the way in which human beings walk is *integration*. This means that we should always strive to integrate the different components to help us gain a deeper insight into the observed gait. Good science should be aimed at emphasizing and explaining underlying causes, rather than merely observing output phenomena, the effects, in some vague and unstructured manner.

Whereas Figures 1.4 and 1.5 show how the different measurements of human gait may be theoretically integrated, Figure 1.6 illustrates how we have implemented this concept in *GaitLab*. The four files on the left—electromyography (EMG), anthropometry (APM), segment kinematics (KIN), and ground reaction data from force plates (FPL)—are all based on direct measurements of the human subject. The other files—body segment parameters (BSP), joint positions and segment endpoints (JNT), reference frames defining segment orientations (REF), centers of gravity, their velocities and

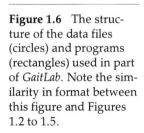

Figure 1.6 The structure of the data files (circles) and programs (rectangles) used in part of *GaitLab*. Note the similarity in format between this figure and Figures 1.2 to 1.5.

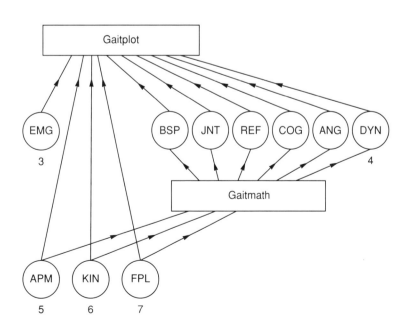

accelerations (COG), joint angles as well as segment angular velocities and accelerations (ANG), dynamic forces and moments at joints (DYN)—are all derived in the program Gaitmath. All 10 files may be viewed and (where appropriate) graphed in the program Gaitplot. The key, emphasized earlier, is integration. Further details on these programs are contained in *GaitLab*.

Summary

This first chapter has given you a framework for understanding how the human body moves. Although the emphasis has been on human gait, the model can be applied in a general way to all types of movement. In the next chapter we introduce you to the basics of human gait, describing its cyclic nature and how we use this periodicity in gait analysis.

CHAPTER 2

The Three-Dimensional and Cyclic Nature of Gait

Most textbooks on anatomy have a diagram, similar to Figure 2.1, that explains the three primary planes of the human body: sagittal, coronal (or frontal), and transverse. Unfortunately, many textbook authors (e.g., Winter, 1987) and researchers emphasize the sagittal plane and ignore the other two. Thus, the three-dimensional nature of human gait has often been overlooked. Although the sagittal plane is probably the most important one, where much of the movement takes place (see Figure 2.2a), there are certain pathologies where another plane (e.g., the coronal, in the case of bilateral hip pain) would yield far more useful information (see Figure 2.2, a-c).

Other textbook authors (Inman et al., 1981; Sutherland, 1984; Sutherland, Olshen, Biden, & Wyatt, 1988) have considered the three-dimensional nature of human gait, but they have looked at the human walker from two or three

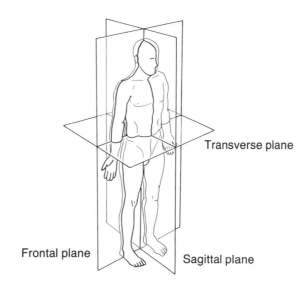

Figure 2.1 The reference planes of the human body in the standard anatomical position. *Note.* From *Human Walking* (p. 34) by V.T. Inman, H.J. Ralston, and F. Todd, 1981, Baltimore: Williams & Wilkins. Copyright 1981 by Williams & Wilkins. Adapted by permission.

Figure 2.2 The gait of an 8-year-old boy as seen in the three principal planes: (a) sagittal; (b) transverse; and (c) frontal.

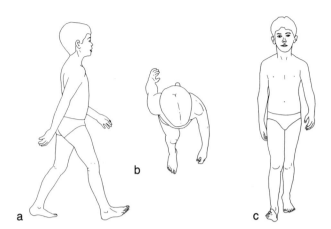

separate views (see Figure 2.3). Though this is clearly an improvement, we believe that the analysis of human gait should be truly three-dimensional: The three separate projections should be combined into a composite image, and the parameters expressed in a body-based rather than laboratory-based coordinate system. This important concept is described further in chapter 3.

Figure 2.3 The walking subject projected onto the three principal planes of movement. *Note.* From *Human Walking* (p. 33) by V.T. Inman, H.J. Ralston, and F. Todd, 1981, Baltimore: Williams & Wilkins. Copyright 1981 by Williams & Wilkins. Adapted by permission.

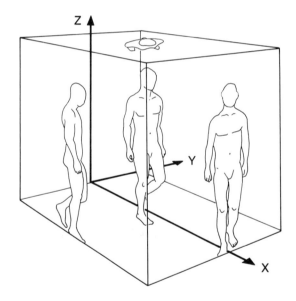

Periodicity of Gait

The act of walking has two basic requisites:

1. Periodic movement of each foot from one position of support to the next
2. Sufficient ground reaction forces, applied through the feet, to support the body

These two elements are necessary for any form of bipedal walking to occur, no matter how distorted the pattern may be by underlying pathology (Inman et al., 1981). This periodic leg movement is the essence of the *cyclic* nature of human gait.

Figure 2.4 illustrates the movement of a wheel from left to right. In the position at which we first see the wheel, the highlighted spoke points vertically down. (The wheel is not stationary here; a "snapshot" has been taken *as the spoke passes through* the vertical position.) By convention, the beginning of the cycle is referred to as 0%. As the wheel continues to move from left to right, the highlighted spoke rotates in a clockwise direction. At 20% it has rotated through 72° (20% × 360°), and for each additional 20%, it advances another 72°. When the spoke returns to its original position (pointing vertically downward), the cycle is complete (this is indicated by 100%).

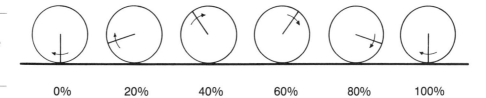

Figure 2.4 A rotating wheel demonstrates the cyclic nature of forward progression.

Gait Cycle

This analogy of a wheel can be applied to human gait. When we think of someone walking, we picture a cyclic pattern of movement that is repeated over and over, step after step. Descriptions of walking are normally confined to a single cycle, with the assumption that successive cycles are all about the same. Although this assumption is not strictly true, it is a reasonable approximation for most people. Figure 2.5 illustrates a single cycle for a normal 8-year-old boy. Note that by convention, the cycle begins when one of the feet (in this case the right foot) makes contact with the ground.

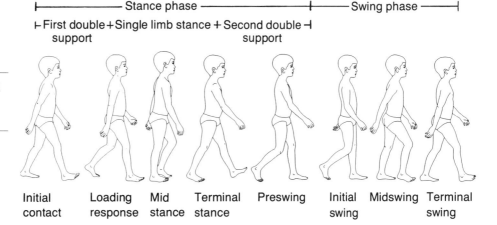

Figure 2.5 The normal gait cycle of an 8-year-old boy.

Phases. There are two main phases in the gait cycle: During *stance* phase, the foot is on the ground, whereas in *swing* phase that same foot is no longer in contact with the ground and the leg is swinging through in preparation for the next foot strike.

As seen in Figure 2.5, the stance phase may be subdivided into three separate phases:

1. *First double support*, when both feet are in contact with the ground

2. *Single limb stance*, when the left foot is swinging through and only the right foot is in ground contact
3. *Second double support*, when both feet are again in ground contact

Note that though the nomenclature in Figure 2.5 refers to the *right* side of the body, the same terminology would be applied to the left side, which for a normal person is half a cycle behind (or ahead of) the right side. Thus, first double support for the right side is second double support for the left side, and vice versa. In normal gait there is a natural symmetry between the left and right sides, but in pathological gait an asymmetrical pattern very often exists. This is graphically illustrated in Figure 2.6. Notice the symmetry in the gait of the normal subject between right and left sides in the stance (62%) and swing (38%) phases; the asymmetry in those phases in the gaits of the two patients, who spend less time bearing weight on their involved (painful) sides; and the increased cycle time for the two patients compared to that of the normal subject.

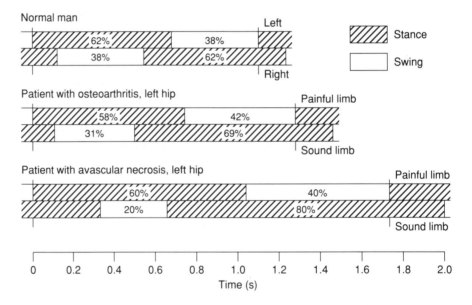

Figure 2.6 The time spent on each limb during the gait cycles of a normal man and two patients with unilateral hip pain. *Note.* Adapted from Murray & Gore (1981).

Events. Traditionally the gait cycle has been divided into eight events or periods, five during stance phase and three during swing. The names of these events are self-descriptive and are based on the movement of the *foot*, as seen in Figure 2.7. In the traditional nomenclature, the stance phase events are as follows:

1. *Heel strike* initiates the gait cycle and represents the point at which the body's center of gravity is at its lowest position.
2. *Foot-flat* is the time when the plantar surface of the foot touches the ground.
3. *Midstance* occurs when the swinging (contralateral) foot passes the stance foot and the body's center of gravity is at its highest position.
4. *Heel-off* occurs as the heel loses contact with the ground and pushoff is initiated via the triceps surae muscles, which plantar flex the ankle.
5. *Toe-off* terminates the stance phase as the foot leaves the ground (Cochran, 1982).

The swing phase events are as follows:

6. *Acceleration* begins as soon as the foot leaves the ground and the subject activates the hip flexor muscles to accelerate the leg forward.
7. *Midswing* occurs when the foot passes directly beneath the body, coincidental with midstance for the other foot.
8. *Deceleration* describes the action of the muscles as they slow the leg and stabilize the foot in preparation for the next heel strike.

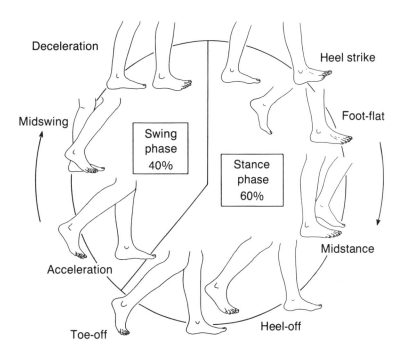

Figure 2.7 The traditional nomenclature for describing eight main events, emphasizing the cyclic nature of human gait.

This traditional nomenclature best describes the gait of normal subjects. However, there are a number of patients with pathologies, such as ankle equinus secondary to spastic cerebral palsy, whose gait cannot be described using this approach. An alternative nomenclature, developed by Perry and her associates at Rancho Los Amigos Hospital in California (Cochran, 1982), is shown in the lower part of Figure 2.5. Here, too, there are eight events, but these are sufficiently general to be applied to any type of gait:

1. Initial contact (0%)
2. Loading response (0-10%)
3. Midstance (10-30%)
4. Terminal stance (30-50%)
5. Preswing (50-60%)
6. Initial swing (60-70%)
7. Midswing (70-85%)
8. Terminal swing (85-100%)

Distance Measures. Whereas Figures 2.5 to 2.7 emphasize the temporal aspects of human gait, Figure 2.8 illustrates how a set of footprints can provide useful distance parameters. *Stride length* is the distance travelled by a person during one stride (or cycle) and can be measured as the length

Figure 2.8 A person's footprints can very often provide useful distance parameters.

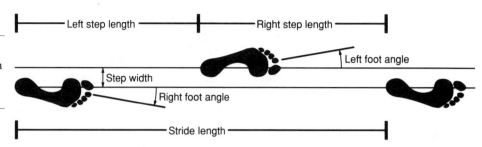

between the heels from one heel strike to the next heel strike on the same side. Two step lengths (left plus right) make one stride length. With normal subjects, the two step lengths will be approximately equal, but with certain patients (such as those illustrated in Figure 2.6), there will be an asymmetry between the left and right sides. Another useful parameter shown in Figure 2.8 is *step width*, which is the mediolateral distance between the feet and has a value of a few centimeters for normal subjects. For patients with balance problems, such as cerebellar ataxia or the athetoid form of cerebral palsy, the stride width can increase to as much as 15 or 20 cm (see the case study in chapter 5). Finally, the angle of the foot relative to the line of progression can also provide useful information, documenting the degree of external or internal rotation of the lower extremity during the stance phase.

Parameters of Gait

The cyclic nature of human gait is a very useful feature for reporting different parameters. As you will later discover in *GaitLab*, there are literally hundreds of parameters that can be expressed in terms of the percent cycle. We have chosen just a few examples (displacement, ground reaction force, and muscle activity) to illustrate this point.

Displacement

Figure 2.9 shows the position of a normal male's right lateral malleolus in the Z (vertical) direction as a function of the cycle. At heel strike, the height is about 0.07 m, and it stays there for the next 40% of the cycle because the

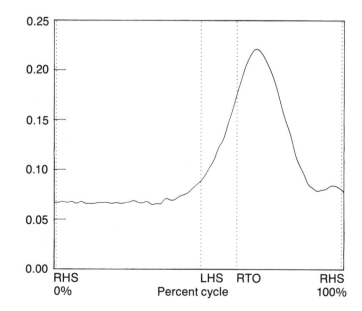

Figure 2.9 The height (value in the Z direction) in meters (m) of a normal male's right malleolus as a function of the gait cycle: RHS, right heel strike; LHS, left heel strike; RTO, right toe-off.

foot is in contact with the ground. Then, as the heel leaves the ground, the malleolus height increases steadily until right toe-off at about 60%, when its height is 0.17 m. After toe-off, the knee continues to flex, and the ankle reaches a maximum height of 0.22 m at 70% of the cycle. Thereafter, the height decreases steadily as the knee extends in preparation for the following right heel strike at 100%. This pattern will be repeated over and over, cycle after cycle, as long as the subject continues to walk on level ground.

Ground Reaction Force

Figure 2.10 shows the vertical ground reaction force of a cerebral palsy adult (whose case is studied in detail in chapter 5) as a function of the gait cycle. Shortly after right heel strike, the force rises to a value over 800 newtons (N) (compared to his weight of about 700 N). By midswing this value has dropped to 400 N, which is a manifestation of his lurching manner of walking. By the beginning of the second double support phase (indicated by LHS, or left heel strike), the vertical force is back up to the level of his body weight. Thereafter it decreases to zero when right toe-off occurs. During the swing phase from right toe-off to right heel strike, the force obviously remains at zero. This ground reaction force pattern is quite similar to that of a normal person except for the exaggerated drop during midstance.

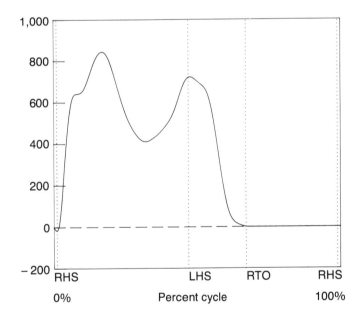

Figure 2.10 The vertical ground reaction force in newtons (N), acting on a cerebral palsy adult's right foot during the gait cycle: RHS, right heel strike; LHS, left heel strike; RTO, right toe-off.

Muscle Activity

Muscle activity, too, can be plotted as a function of percent cycle as seen in Figure 2.11. Here the EMG of the rectus femoris for a normal female is illustrated. Notice that just after right heel strike, the EMG increases. Because the rectus femoris is a hip flexor and knee extensor, but the hip and knee are extending and flexing at this time, the muscle is acting eccentrically. During the midstance phase, the activity decreases substantially, picking up again during late stance and early swing. During this period, both the hip and knee are flexing. The rectus femoris is again reasonably quiescent in midswing, but its activity increases before the second right heel strike.

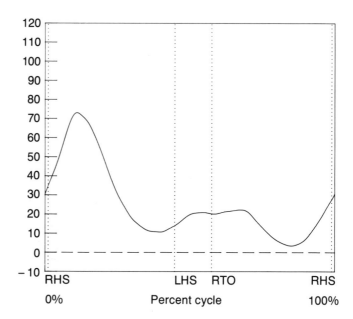

Figure 2.11 The electromyographic activity in microvolts (μV) of a normal female's right rectus femoris muscle plotted as a function of the gait cycle: RHS, right heel strike; LHS, left heel strike; RTO, right toe-off. *Note.* Data from Winter (1987).

Summary

Figure 2.11 provides some insight into the actions of a single muscle, but literally hundreds of muscles are active during the gait cycle. The challenge facing the central nervous system is to control simultaneously the actions of all these muscles. This is addressed further in chapter 4. Before that, however, chapter 3 teaches you how to integrate anthropometric, kinematic, and force plate data.

CHAPTER 3

Integration of Anthropometry, Displacements, and Ground Reaction Forces

In chapter 1 you learned that the gait analyst must pursue the inverse dynamics approach in which the motion of the mechanical system is completely specified and the objective is to find the forces causing that motion. This approach is illustrated in words in Figure 1.4 and in mathematical symbols in Figure 1.5. In chapter 2 you were introduced to the three-dimensional nature of human gait. You also learned that gait is a cyclic activity and that many variables, such as displacement, ground reaction forces, and muscle activity, can be plotted as a function of the cycle. In this chapter we will show how all these measurements may be integrated to yield the resultant forces and moments acting at the joints of the lower extremities.

This chapter covers five different topics. In Body Segment Parameters, you will learn how simple anthropometric measurements, such as total body mass and calf length, can be used in regression equations to predict the masses and moments of inertia of lower extremity segments. In Linear Kinematics we show how the position of external markers attached to the skin may be used to predict the position of internal landmarks such as the joint centers. In Centers of Gravity, the joint centers are used to predict the positions of the segment centers of gravity; then, using numerical differentiation, the velocities and accelerations of these positions are obtained. In Angular Kinematics, the anatomical joint angles are calculated, as are the angular velocities and accelerations of the segments. Finally, in Dynamics of Joints, the body segment parameters, linear kinematics, centers of gravity, angular kinematics, and ground reaction forces are all integrated in the equations of motion (see Figures 1.4 and 1.5) to yield the resultant joint forces and moments.

Be aware that because we are dealing with gait analysis as a three-dimensional phenomenon, some of the concepts and mathematics are quite complex. However, our intent is that the material in this chapter be accessible to all persons who have had a basic undergraduate course in mathematics. The challenge will be thinking in 3-D even though the diagrams are obviously plotted on a flat, 2-D page. If you need a bigger challenge, a detailed and rigorous coverage of the material is presented in Appendix B.

Body Segment Parameters

A major concern for the gait analyst is personalizing the body segment parameters of the individual subject. By *body segment parameters* we mean

- mass in kilograms of the individual segments (e.g., thigh, calf, foot);
- center of gravity location of the individual segments relative to some specified anatomical landmarks (e.g., proximal and distal joints); and
- moments of inertia of the segments about three orthogonal axes (i.e., axes at right angles to one another) that pass through the segment center of gravity.

Moment of inertia is a measure of the way in which the mass is distributed about the axis of interest and has the units of kilogram • meter • meter (kg • m^2). It therefore varies with mass and the square of length.

Selection of Segments

Another concern is the selection of individual segments. As you will see a little later in the chapter, we have chosen six segments: thigh, calf, and foot on both the left and right sides. We are making the assumption that these are rigid segments whose dimensions (and thus their segment parameters) do not change during the motion of interest. We all know, however, that the foot is not a single rigid segment and so you should be aware that any model has some limitations. We chose a 6-segment model for simplicity (and because virtually all gait laboratories do the same); but it is possible that in the future, biomechanical models will need to be more detailed.

Problems in Estimation

In attempting to estimate the body segment parameters for an individual subject, there are various approaches that can be followed. These include

- cadaver averages (Braune & Fischer, 1889; Dempster, 1955);
- reaction board (Bernstein, 1967);
- mathematical modeling (Hanavan, 1964; Hatze, 1980);
- scans using gamma rays, axial tomography, or magnetic resonance imaging (Brooks & Jacobs, 1975; Erdmann, 1989; Huang & Wu, 1976; Martin, Mungiole, Marzke, & Longhill, 1989; Zatsiorsky & Seluyanov, 1985); and
- kinematic measurements (Ackland, Blanksby, & Bloomfield, 1988; Dainis, 1980; Vaughan, Andrews, & Hay, 1982).

Each of these has severe limitations. The cadaver averages are not sufficiently specific for individual subjects and very often only total body mass is used as a predictive variable. The reaction board technique is a long and tedious procedure which cannot estimate segment masses and centers of gravity independently. Mathematical modeling suffers from the disadvantage that too many variables (242 in the case of Hatze's model) need to be measured, thus requiring an inordinate amount of time and patience. Scanning techniques, though potentially very accurate and detailed, must be seriously questioned as a routine method because of the radiation exposure and high costs. Although they have some appeal, kinematic measurements either have not yielded results to a satisfactory degree of accuracy or require too much time (Jensen, 1986).

Anthropometry

What is needed for estimating body segment parameters is a technique with the following features:

- Personalized for individuals
- Short time required to take measurements
- Inexpensive and safe
- Reasonably accurate

We can describe a technique that we believe meets these criteria. Not surprisingly, it is based on anthropometry. Figure 3.1 illustrates the measurements that need to be made, Table 3.1 describes, in anatomical terminology, how the parameters are measured, and Table 3.2 shows the data for a normal man. (In fact, Table 3.2 contains the data for the Man.APM file used in Gaitmath.)

There are 20 measurements that need to be taken—9 for each side of the body, plus the subject's total body mass, and the distance between the anterior superior iliac spines (ASIS). With experience, these measurements can be made in less than 10 min using standard tape measures and beam calipers, which are readily available. They describe, in some detail, the characteristics of the subject's lower extremities. The question to be answered is this: Can they be used to predict body segment parameters that are specific to the individual subject and reasonably accurate? We believe the answer is yes.

As mentioned earlier, most of the regression equations based on cadaver data use only total body mass to predict individual segment masses. Although this will obviously provide a reasonable estimate as a first approximation, it does not take into account the variation in the shape of the individual segments.

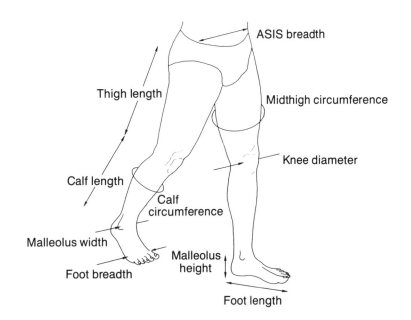

Figure 3.1 The anthropometric measurements of the lower extremity that are required for the prediction of body segment parameters (masses and moments of inertia).

Table 3.1 Description of Anthropometric Parameters and How to Measure Them

Parameter	Description
Body mass	Measure (on a scale accurate to 0.01 kg) the mass of subject with all clothes except underwear removed
ASIS breadth	With a beam caliper, measure the horizontal distance between the anterior superior iliac spines
Thigh length	With a sliding caliper, measure the vertical distance between the superior point of the greater trochanter of the femur and the superior margin of the lateral tibia
Midthigh circumference	With a tape perpendicular to the long axis of the leg and at a level midway between the trochanteric and tibial landmarks, measure the circumference of the thigh
Calf length	With a sliding caliper, measure the vertical distance between the superior margin of the lateral tibia and the lateral malleolus
Calf circumference	With a tape perpendicular to the long axis of the lower leg, measure the maximum circumference of the calf
Knee diameter	With a spreading caliper, measure the maximum breadth of the knee across the femoral epicondyles
Foot length	With a beam caliper, measure the distance from the posterior margin of the heel to the tip of the longest toe
Malleolus height	With the subject standing, use a sliding caliper to measure the vertical distance from the standing surface to the lateral malleolus
Malleolus width	With a sliding caliper, measure the maximum distance between the medial and lateral malleoli
Foot breadth	With a beam caliper, measure the breadth across the distal ends of metatarsals I and V

Note. Adapted from Chandler et al. (1975).

Prediction of Segment Mass

We believe that individual segment masses are related not only to the subject's total body mass, but also to the dimensions of the segment of interest. Specifically, because mass is equal to density times volume, the segment mass should be related to a composite parameter which has the dimensions of length cubed and depends on the volume of the segment. Expressed mathematically, we are seeking a multiple linear regression equation for predicting segment mass which has the form

$$\text{Segment mass} = C1\,(\text{Total body mass}) + C2\,(\text{Length})^3 + C3 \quad (3.1)$$

where C1, C2, and C3 are regression coefficients. For our purposes, the shapes of the thigh and calf are represented by cylinders, and the shape of the foot is similar to a right pyramid. (Though a truncated cone may provide a better approximation of the thigh and calf, it requires an extra anthropometric measurement for each segment and yields only a marginal improvement in accuracy.)

$$\text{Mass of cylinder} = \frac{(\text{Density})}{4\pi}(\text{Length})(\text{Circumference})^2 \quad (3.2)$$

Table 3.2 **Anthropometric Data Required to Predict Body Segment Parameters for a Normal Male**

Number	Anthropometric measurement	Value	Units
1	Total body mass	64.90	kg
2	ASIS breadth	0.240	m
3	R. Thigh length	0.460	m
4	L. Thigh length	0.465	m
5	R. Midthigh circumference	0.450	m
6	L. Midthigh circumference	0.440	m
7	R. Calf length	0.430	m
8	L. Calf length	0.430	m
9	R. Calf circumference	0.365	m
10	L. Calf circumference	0.365	m
11	R. Knee diameter	0.108	m
12	L. Knee diameter	0.112	m
13	R. Foot length	0.260	m
14	L. Foot length	0.260	m
15	R. Malleolus height	0.060	m
16	L. Malleolus height	0.060	m
17	R. Malleolus width	0.074	m
18	L. Malleolus width	0.073	m
19	R. Foot breadth	0.098	m
20	L. Foot breadth	0.096	m

$$\text{Mass of pyramid} = \frac{(\text{Density})}{3} (\text{Width})(\text{Height})(\text{Length}) \qquad (3.3)$$

Figure 3.2, a-c, illustrates the three anatomical segments and their geometric counterparts.

We assume the segment density among subjects to be invariant and then use the linear dimensions as predictors of the segment masses. We based our regression equations on six cadavers studied by Chandler, Clauser, McConville, Reynolds, and Young (1975). Although we would ideally prefer to have had more cadavers, these are the only data in the literature that are so complete. This applies not only to the anthropometric measurements (Tables 3.1 and 3.2) but also to the principal centroidal moments of inertia which are described in great detail. Based on these data in Chandler et al.

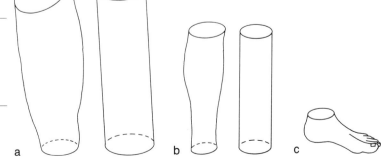

Figure 3.2 Lower extremity body segments and their geometric counterparts: (a) thigh; (b) calf; (c) foot.

(1975), we have performed multiple linear regression to yield the following equations:

$$\begin{aligned}\text{Mass of thigh} = &\ (0.1032)\,(\text{Total body mass}) \\ &+ (12.76)\,(\text{Thigh length})\,(\text{Midthigh circumference})^2 \\ &+ (-1.023)\end{aligned}$$

(3.4)

$$\begin{aligned}\text{Mass of calf} = &\ (0.0226)\,(\text{Total body mass}) \\ &+ (31.33)\,(\text{Calf length})\,(\text{Calf circumference})^2 \\ &+ (0.016)\end{aligned}$$

(3.5)

$$\begin{aligned}\text{Mass of foot} = &\ (0.0083)\,(\text{Total body mass}) \\ &+ (254.5)\,(\text{Malleolus width})\,(\text{Malleolus height}) \\ &\ (\text{Foot length}) + (-0.065)\end{aligned}$$

(3.6)

Using the data in Table 3.2 (Total body mass = 64.90 kg; Right calf length = 0.430 m; Right calf circumference = 0.365 m), Equation 3.5 yields

$$\begin{aligned}\text{Mass of right calf} &= (0.0226)\,(64.90) \\ &\quad + (31.33)\,(0.430)\,(0.365)^2 + 0.016 \\ &= 3.28 \text{ kg}\end{aligned}$$

Further details on the statistical basis for these regression equations to predict mass may be found in Appendix B.

Prediction of Segment Moments of Inertia

As mentioned previously, the moment of inertia, which is a measure of a body's resistance to angular motion, has units of kgm^2. It seems likely therefore that the moment of inertia would be related to body mass (kilogram) times a composite parameter which has the dimensions of length squared (m^2). Expressed mathematically, we are seeking a linear regression equation for predicting segment moment of inertia which has the form

$$\text{Segment moment of inertia} = C4\,(\text{Total body mass})\,(\text{Length})^2 + C5$$

(3.7)

where C4 and C5 are regression coefficients. The key is to recognize that the (Length)^2 parameter is based on the moment of inertia of a similarly shaped, geometric solid. A similar approach has recently been proposed by Yeadon and Morlock (1989). As before, the thigh and calf are similar to a cylinder and the foot is approximated by a right pyramid. Figure 3.3 shows the principal orthogonal axes for the thigh and a cylinder.

Using the mathematical definition of moment of inertia and standard calculus, the following relationships can be derived:

Moment of inertia of cylinder about the flexion/extension axis =

$$\frac{(\text{Mass})}{12}\,[(\text{Length})^2 + 0.076\,(\text{Circumference})^2]$$

(3.8)

Moment of inertia of cylinder about the abduction/adduction axis =

$$\frac{(\text{Mass})}{12}\,[(\text{Length})^2 + 0.076\,(\text{Circumference})^2]$$

(3.9)

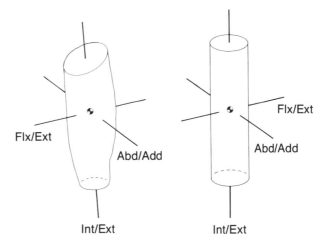

Figure 3.3 Principal orthogonal axes for the thigh and a right cylinder.

Moment of inertia of cylinder about the internal/external rotation axis =

$$\frac{(\text{Mass})}{8\pi^2}(\text{Circumference})^2$$

(3.10)

When studying these three equations, you will notice the following: Equations 3.8 and 3.9 are the same, which comes from the radial symmetry of a cylinder; all three equations have the units of kg · m^2; the moment of inertia about the internal/external rotation axis (Equation 3.10) depends only on mass and circumference (squared) and not on the length of the cylinder. As can be seen from Equation 3.7, each of the equations provides two regression coefficients. There are three regression equations per segment (Equations 3.8, 3.9, and 3.10 are the examples for the thigh), and there are three separate segments—thigh, calf, and foot. This means that the regression analysis of the Chandler data will yield 2 × 3 × 3 = 18 regression coefficients. All of these are provided in Appendix B, but for the purpose of this chapter, we show one regression equation for the thigh:

Moment of inertia of thigh about the flexion/extension axis =
(0.00762) (Total body mass) ×
[(Thigh length)^2 + 0.076 (Midthigh circumference)^2] + 0.0115

(3.11)

Again, using the data in Table 3.2 (Total body mass = 64.90 kg; Right thigh length = 0.460 m; Right midthigh circumference = 0.450 m) in the previous equation, we get

Moment of inertia of right thigh about the flexion/extension axis =
(0.00762) (64.90) × [(0.460)^2 + 0.076 (0.450)^2] + 0.0115 = 0.1238 kg · m^2

(3.12)

Table 3.3, which contains the data for the Man.BSP file generated in Gaitmath, provides all the body segment parameters that are required for detailed 3-D gait analysis of the lower extremities. In addition to the body segment masses and moments of inertia already discussed in this section, notice that there are also segment center-of-mass data. These are expressed as ratios and are based on knowing the segment endpoints for the thigh, calf, and foot. These points are between the hip and knee joints, the knee and ankle joints, and the heel and longest toe, respectively. The ratios in Table 3.3 are the means of the ratios in Chandler et al. (1975).

Table 3.3 Body Segment Parameter Data for Lower Extremities of a Normal Male

Segment number Segment name	1 R. Thigh	2 L. Thigh	3 R. Calf	4 L. Calf	5 R. Foot	6 L. Foot
Mass (kg)	6.86	6.82	3.28	3.28	0.77	0.76
CG position (ratio: proximal/length)	0.39	0.39	0.42	0.42	0.44	0.44
Moments of inertia (kg•m•m)						
Flx/Ext axis	0.1238	0.1257	0.0490	0.0490	0.0035	0.0035
Abd/Add axis	0.1188	0.1207	0.0504	0.0504	0.0040	0.0039
Int/Ext axis	0.0229	0.0220	0.0037	0.0037	0.0011	0.0011

Note. The parameters were generated from the anthropometric data of Table 3.2 using regression equations such as 3.4 to 3.6 and 3.11.

In summary then, Table 3.1 describes how the anthropometric measurements should be made, Table 3.2 is an example of the 20 parameters for a male subject, and Table 3.3 shows the body segment parameters (BSPs) that are derived using the regression equations and anthropometric measurements. We think you will agree that the BSPs have been personalized by means of linear measurements that do not require much time or expensive equipment. In Appendix B, we show that these equations are also reasonably accurate and can therefore be used with some confidence.

Though we believe that our BSPs are superior to other regression equations that are *not* dimensionally consistent (Hinrichs, 1985), it is appropriate to put this statement into the proper perspective. The moments of inertia are really only needed to calculate the resultant joint moments (see Equation 3.30 later in this chapter). Their contribution is relatively small, particularly for the internal/external rotation axis. For example, in stance phase, the contributions from the inertial terms to joint moments are very small because the velocity and acceleration of limb segments are small.

Linear Kinematics

As described in the previous section on anthropometry, each of the segments of the lower extremity (thigh, calf, and foot) may be considered as a separate entity. Modeling the human body as a series of interconnected rigid links is a standard biomechanical approach (Apkarian, Naumann, & Cairns, 1989; Capozzo, 1984). When studying the movement of a segment in 3-D space we need to realize that it has six degrees of freedom. This simply means that it requires six independent coordinates to describe its position in 3-D space uniquely (Greenwood, 1965). You may think of these six as being three cartesian coordinates (X, Y, and Z) and three angles of rotation, often referred to as Euler angles. In order for the gait analyst to derive these six coordinates, he or she needs to measure the 3-D positions of at least three noncolinear markers on each segment. The question that now arises is this: Where on the lower extremities should these markers be placed? Ideally, we want the minimum number of markers placed on anatomical landmarks that can be reliably located, otherwise data capture becomes tedious and prone to errors.

Use of Markers

Some systems, such as the commercially available OrthoTrak product from Motion Analysis (see Appendix C), use up to 25 markers. We feel this is too many markers, and the use of bulky triads on each thigh and calf severely encumbers the subject. Kadaba, Ramakrishnan, and Wootten (1990) have proposed a 15-marker system that uses wands or sticks about 7 to 10 cm long attached to the pelvis, thighs, and calves. The advantage of this approach is that the markers are easier to track in 3-D space with video-based kinematic systems, and they can (at least theoretically) provide more accurate orientation of the segment in 3-D space. The major disadvantages are that the wands encumber the subject, and if he or she has a jerky gait, the wands will vibrate and move relative to the underlying skeleton. In addition, Kadaba et al. (1990) only have two markers on the foot segment. After careful consideration we have chosen the 15 marker locations illustrated in Figure 3.4, a and b.

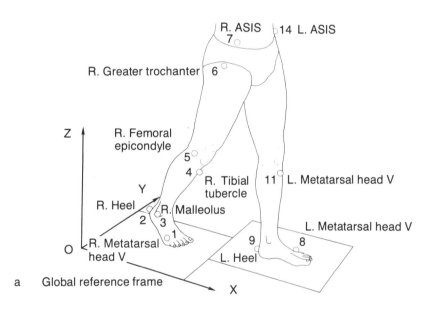

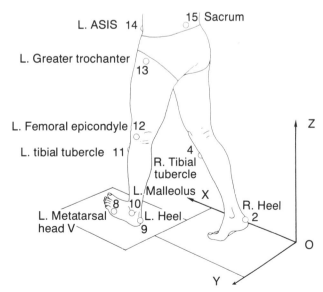

Figure 3.4 The 15-marker system that uniquely defines the position of each segment in 3-D space: (a) anterior view; (b) posterior view.

Note the position of the XYZ global reference system in Figure 3.4 with its origin at one corner of Force Plate 1. The X, Y, and Z coordinates of these 15 markers as a function of time may then be captured with standard equipment (described in Appendix C). The data in the KIN files in *GaitLab* were gathered at the Oxford Orthopaedic Engineering Centre (OOEC), the National Institutes of Health (NIH) Biomechanics Laboratory, and the Richmond Children's Hospital (where all laboratories use the VICON system from Oxford Metrics). Table 3.4 shows the data for one time frame (actually the first right heel strike), while Figure 3.5 is an example of X, Y, and Z coordinates of the right lateral malleolus plotted as a function of time.

Table 3.4 Three-Dimensional Displacement Data of External Landmarks at Time = 0.00 s (Right Heel Strike) for a Normal Male

Number	Name	X (m)	Y (m)	Z (m)
1	R. Metatarsal head V	0.157	0.142	0.057
2	R. Heel	−0.016	0.207	0.032
3	R. Lateral malleolus	0.033	0.155	0.067
4	R. Tibial tubercle	0.002	0.181	0.443
5	R. Femoral epicondyle	−0.055	0.136	0.509
6	R. Greater trochanter	−0.207	0.082	0.880
7	R. ASIS	−0.161	0.141	0.979
8	L. Metatarsal head V	−0.578	0.392	0.032
9	L. Heel	−0.705	0.320	0.138
10	L. Lateral malleolus	−0.648	0.374	0.128
11	L. Tibial tubercle	−0.383	0.341	0.396
12	L. Femoral epicondyle	−0.369	0.393	0.495
13	L. Greater trochanter	−0.263	0.439	0.891
14	L. ASIS	−0.174	0.386	0.976
15	Sacrum	−0.369	0.242	0.962

Note. The XYZ positions refer to the global coordinate system defined in Figure 3.4, although the position of the subject in this figure is much later in the gait cycle (approaching right toe-off).

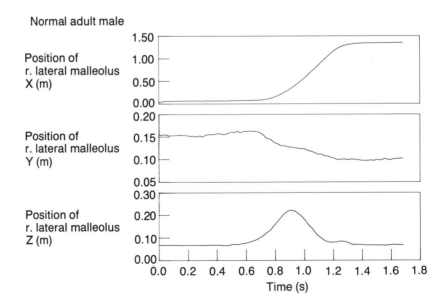

Figure 3.5 The 3-D coordinates of the right lateral malleolus plotted as a function of time. There are approximately one and a half gait cycles (i.e., from right heel strike to beyond the next right heel strike) in this figure, as can be seen in the Z curve.

Marker Placement for Current Model

One of the problems in capturing kinematic data is that we are really interested in the position of the underlying skeleton, but we are only able to measure the positions of external landmarks (Figure 3.4 and Table 3.4). Because most gait studies are two-dimensional and concentrate on the sagittal plane, researchers have assumed that the skeletal structure of interest lies behind the external marker. We obviously cannot do that with our 3-D marker positions, but we can use the external landmarks to predict internal positions. The 3-step strategy used to calculate the positions of the hip, knee, and ankle joints on both sides of the body is as follows:

1. Select three markers for the segment of interest.
2. Create an orthogonal **uvw** reference system based on these three markers.
3. Use prediction equations based on anthropometric measurements and the **uvw** reference system to estimate the joint center positions.

Foot. Consider the markers on the right foot as seen in Figure 3.4a. These are numbered 1, metatarsal head V; 2, heel; and 3, lateral malleolus. They are shown in more detail in Figure 3.6, a and b.

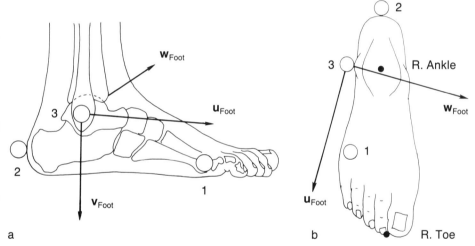

Figure 3.6 The three markers (1, 2, and 3) which define the position of the foot in 3-D space: (a) side view; (b) view from above. The **uvw** reference system may be used to predict the position of the ankle and toe.

When creating the **uvw** reference system, we first place the origin at Marker 3 (lateral malleolus). The three markers form a plane, and the **w** axis is perpendicular to this plane. The **u** axis is parallel to the line between Markers 2 and 1, although its origin is Marker 3. Finally, the **v** axis is at right angles to both **u** and **w** so that the three axes **uvw** form a so-called right-handed system. (To determine if a system is right-handed, point the fingers of your right hand in the direction of the **u** axis, curl them toward the **v** axis, and your thumb should be pointing in the **w** direction. This is called the right-handed screw rule.) Now that **uvw** for the foot has been defined, we can use this information in prediction equations to estimate the position of the ankle and longest toe:

$$\mathbf{p}_{\text{Ankle}} = \mathbf{p}_{\text{Lateral malleolus}} \\ - (0.008)(\text{Foot length})\,\mathbf{u}_{\text{Foot}} \\ + (0.393)(\text{Malleolus height})\,\mathbf{v}_{\text{Foot}} \\ + (0.706)(\text{Malleolus width})\,\mathbf{w}_{\text{Foot}} \quad (3.13)$$

$$\begin{aligned}\mathbf{p}_{\text{Toe}} = \mathbf{p}_{\text{Lateral malleolus}} & \\ + (0.697)\,(\text{Foot length})\,\mathbf{u}_{\text{Foot}} & \\ + (0.780)\,(\text{Malleolus height})\,\mathbf{v}_{\text{Foot}} & \\ + (0.923)\,(\text{Foot breadth})\,\mathbf{w}_{\text{Foot}} & \end{aligned} \quad (3.14)$$

You should realize that these equations refer to the *right* ankle and toe. The mathematics for calculating **uvw** and distinguishing between the left and right sides may be found in Appendix B.

Calf. Consider the markers on the right calf as seen in Figure 3.4a (note: This segment is sometimes referred to as the shank or leg). These are numbered 3, lateral malleolus; 4, tibial turbercle; and 5, femoral epicondyle. They appear in more detail in Figure 3.7.

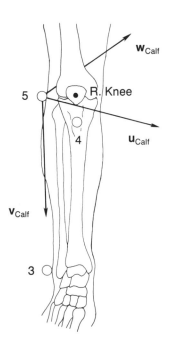

Figure 3.7 The three markers (3, 4, and 5), which define the position of the calf in 3-D space. This is an anterior view. The **uvw** reference system may be used to predict the position of the knee joint.

When creating the **uvw** reference system, we first place the origin at Marker 5, femoral epicondyle. The three markers form a plane, and the **w** axis is perpendicular to this plane. The **v** axis is parallel to the line between Markers 5 and 3. Finally, the **u** axis is at right angles to both **v** and **w** so that the three axes **uvw** form a right-handed system as before. We can now use this triad **uvw** for the calf to estimate the position of the knee joint center based on the following prediction equation:

$$\begin{aligned}\mathbf{p}_{\text{Knee}} = \mathbf{p}_{\text{Femoral epicondyle}} & \\ + (0.423)\,(\text{Knee diameter})\,\mathbf{u}_{\text{Calf}} & \\ - (0.198)\,(\text{Knee diameter})\,\mathbf{v}_{\text{Calf}} & \\ + (0.406)\,(\text{Knee diameter})\,\mathbf{w}_{\text{Calf}} & \end{aligned} \quad (3.15)$$

As with the ankle, this equation refers to the right knee, but the mathematics for the left knee are essentially the same (see Appendix B).

Pelvis. Consider the markers on the pelvis as seen in Figure 3.4. These are numbered 7, right anterior superior iliac spine or ASIS; 14, left ASIS; and 15, sacrum. The sacral marker is placed at the junction between the fifth lumbar vertebra and the sacrum. They appear in more detail in Figure 3.8, a and b.

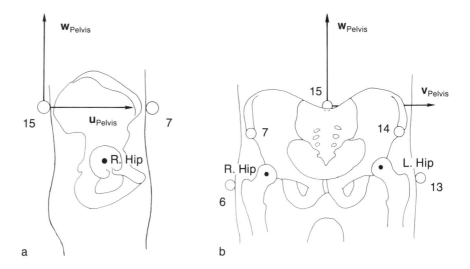

Figure 3.8 The three markers (7, 14, and 15), which define the position of the pelvis in 3-D space: (a) lateral view; (b) anterior view. The **uvw** reference system may be used to predict the position of the right and left hips.

When creating the **uvw** reference system for the pelvis, we first place the origin at Marker 15 (sacrum). The three markers form a plane, and the **w** axis is perpendicular to this plane. The **v** axis is parallel to the line between Markers 7 and 14, although its origin is Marker 15. Finally, the **u** axis is at right angles to both **v** and **w** so that the three axes **uvw** form a right-handed system. Now that **uvw** for the pelvis has been defined, we can use this information in a prediction equation to estimate the positions of the left and right hip joints:

$$\begin{aligned} \mathbf{p}_{Hip} = \;& \mathbf{p}_{Sacrum} \\ & + (0.598)\,(\text{ASIS breadth})\,\mathbf{u}_{Pelvis} \\ & \pm (0.344)\,(\text{ASIS breadth})\,\mathbf{v}_{Pelvis} \\ & - (0.290)\,(\text{ASIS breadth})\,\mathbf{w}_{Pelvis} \end{aligned} \qquad (3.16)$$

The ± differentiates between the left (+) and right (−) hip joints. This method for predicting the positions of the hip joint centers is very similar to others in the literature (Campbell, Grabiner, Hawthorne, & Hawkins, 1988; Tylkowski, Simon, & Mansour, 1982).

Prediction of Joint Centers

Equations 3.13 to 3.16 demonstrate how it is possible to use the 3-D positions of external landmarks (Table 3.4) and anthropometric data (Table 3.2) to predict the 3-D positions of internal skeletal landmarks (i.e., the joint centers). All the coefficients in Equations 3.13 to 3.16 have been based on stereo X rays of one normal subject (Vaughan, 1983). The biomechanics literature desperately needs coefficients that have been derived from many subjects, both normal and abnormal. Table 3.5 shows an example of the 3-D data of the joint centers and foot endpoints from the Man.JNT file in *GaitLab*. A single joint, the right ankle, has been plotted out in Figure 3.9 for comparison with the right lateral malleolus in Figure 3.5. Though the curves are similar, they are not identical, especially with regard to the absolute value of the Y coordinate.

Determination of Segment Orientation

The final task in this section on linear kinematics is to determine the orientation of each segment in 3-D space. This is done by embedding a reference

Table 3.5 Three-Dimensional Displacement Data of the Joint Centers and Foot Endpoints at Time = 0.00 s (Right Heel Strike) in a Normal Male

Joint/Point	X (m)	Y (m)	Z (m)
R. Hip	−0.193	0.161	0.905
L. Hip	−0.201	0.326	0.901
R. Knee	−0.037	0.197	0.534
L. Knee	−0.327	0.326	0.509
R. Ankle	0.059	0.204	0.090
L. Ankle	−0.610	0.326	0.138
R. Heel	−0.007	0.205	0.036
L. Heel	−0.695	0.321	0.152
R. Toe	0.261	0.232	0.133
L. Toe	−0.441	0.308	0.018

Note. The XYZ positions refer to the global coordinate system defined in Figure 3.4.

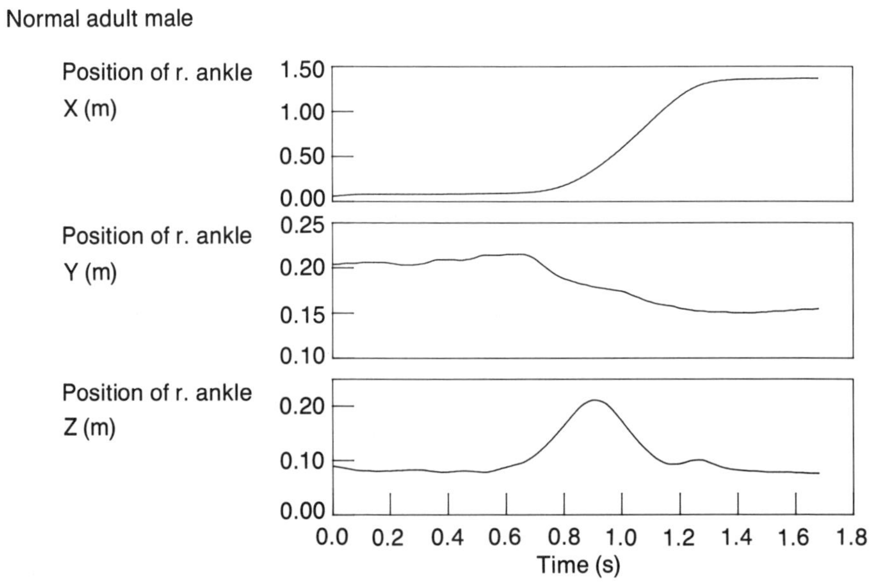

Figure 3.9 The 3-D coordinates of the right ankle joint while walking, plotted as a function of time. Note the similarities (and differences) between these curves and those in Figure 3.5, which are for the right lateral malleolus.

system (xyz) in each segment that will define how each segment is positioned relative to the global (i.e., laboratory) reference frame XYZ. The location of each xyz reference frame is at the segment's center of gravity. (We will describe in the next section how this position is obtained.) Figure 3.10 illustrates how the xyz frames are derived.

In the case of the thighs, the x axis runs from distal to proximal on a line between the knee and hip joints. The xz plane is formed by the hip joint, the greater trochanter marker, and the knee joint. The y axis, by definition, is at right angles to the xz plane and, as seen in Figure 3.10, points in an anterior direction. By the right-handed screw rule (defined earlier), the z axis is perpendicular to x and y and points to the person's left.

For the calves, the x axis again runs from distal to proximal but on a line between the ankle and knee joints. The xz plane is formed by the knee joint, the lateral femoral epicondyle marker, and the ankle joint. The y axis, by definition, is at right angles to the xz plane and, as seen in Figure 3.10, points in an anterior direction. Using the right-handed screw rule again, the z axis

Figure 3.10 The segment reference frames (xyz) embedded at the centers of gravity of each segment. Note the segment numbering system: 1, right thigh; 2, left thigh; 3, right calf; 4, left calf; 5, right foot; and 6, left foot. Refer to the text for the definition of these reference frames.

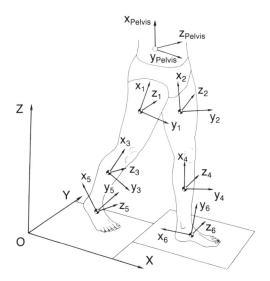

is perpendicular to x and y and points to the person's left. For the feet, the x axis is directed from the longest toe toward the heel marker. The xy plane is formed by the ankle joint, the heel marker, and the toe position. The z axis, by definition, is at right angles to the xy plane and points to the person's left. As seen in Figure 3.10, the y axis obtained by the right-handed screw rule (in this case $z \rightarrow x \rightarrow y$) points upward and away from the dorsal surface of the foot.

The mathematics to determine xyz for each segment relative to the global XYZ is based on standard vector algebra and is covered in full detail in Appendix B.

Centers of Gravity

In order to calculate the positions of segment centers of gravity (CG), we need two sets of information: the locations of the joint centers or segment endpoints (Table 3.5) and the body segment parameter data (Table 3.3).

Prediction of Segment Center of Gravity

Using the mean data of Chandler et al. (1975) and the example shown in Figure 3.11, the following equations for predicting center of gravity locations may be derived:

$$\mathbf{p}_{Thigh.CG} = \mathbf{p}_{Hip} + 0.39\,(\mathbf{p}_{Knee} - \mathbf{p}_{Hip}) \quad (3.17)$$

$$\mathbf{p}_{Calf.CG} = \mathbf{p}_{Knee} + 0.42\,(\mathbf{p}_{Ankle} - \mathbf{p}_{Knee}) \quad (3.18)$$

$$\mathbf{p}_{Foot.CG} = \mathbf{p}_{Heel} + 0.44\,(\mathbf{p}_{Toe} - \mathbf{p}_{Heel}) \quad (3.19)$$

Velocity and Acceleration

The next task is to get the velocity and acceleration data for the segment centers of gravity (refer to Figures 1.4 and 1.5). Velocity and acceleration may be defined, in simplistic form, as follows:

$$\text{Velocity} = \frac{\text{(Change in displacement)}}{\text{(Change in time)}} \quad (3.20)$$

$$\text{Acceleration} = \frac{\text{(Change in velocity)}}{\text{(Change in time)}} \quad (3.21)$$

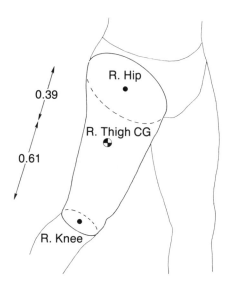

Figure 3.11 The location of the center of gravity of the right thigh, with the joint centers (hip and knee) and corresponding body segment parameters (0.39, 0.61).

These two equations are simple, finite difference equations and are part of the branch of mathematics known as numerical differentiation (Miller & Nelson, 1973). Two techniques for smoothing out noisy displacement data and performing numerical differentiation were presented by Vaughan (1982): a digital filter and the quintic spline. Though the quintic spline can be used to good effect for both smoothing and differentiation (Wood & Jennings, 1979), the size of the computer code (i.e., the number of program lines) and its long running time precludes its use on a personal computer. For *GaitLab*, we have chosen to use a digital filter algorithm for smoothing displacement data, as popularized for use in biomechanics research by Pezzack, Norman, and Winter (1977).

Example of Center of Gravity Data

Table 3.6 is an example of the center of gravity data for one time frame for the Man.COG file in *GaitLab*. Choosing the center of gravity (CG) of one segment (the right foot) and one direction (Z) yields the curves seen in Figure 3.12. This figure shows how Equations 3.20 and 3.21 function. Study the position curve at the top of Figure 3.12 first. You will notice that at $t = 0.0$ s (heel strike), the foot's CG is slightly above zero. It drops down almost to zero and then does not change much until about 0.6 s. This period is obviously when the foot is flat on the ground. At heel strike, the velocity is slightly negative because the CG is still moving downward. During foot-flat (0.1-0.6 s), the velocity is almost zero because the displacement is not changing (see Equation 3.20). The acceleration is quite large and positive just after heel strike when the foot is experiencing an upward force from the ground. During the foot-flat phase, when the velocity is not changing much, the acceleration is almost zero (see Equation 3.21).

The next phase is heel-off to toe-off which takes place between 0.6 and 0.8 s. The position of the foot's CG gradually increases to about 0.15 m. The velocity increases to 0.7 m/s as the heel lifts off prior to toe-off at 0.8 s. From toe-off (just after 0.8 s) to the next heel strike (just after 1.3 s), there are some interesting hills and valleys in all three curves. From Equation 3.20, you can see that any time the position curve has a hill or valley (i.e., a maximum or minimum point), the velocity must be zero. Similarly, from Equation 3.21, any time the velocity curve has a hill or valley (also referred to as a *turning*

Table 3.6 Three-Dimensional Displacement, Velocity, and Acceleration Data of the Segment Centers of Gravity at Time = 0.00 s (Right Heel Strike) for a Normal Male

Segment	X	Y	Z	v_X	v_Y	v_Z	a_X	a_Y	a_Z
R. Thigh	−0.121	0.173	0.761	0.68	−0.12	0.08	8.2	−1.3	2.6
L. Thigh	−0.238	0.324	0.748	0.77	−0.09	−0.02	8.7	0.5	0.3
R. Calf	0.012	0.199	0.347	0.52	−0.05	0.00	4.0	0.6	1.4
L. Calf	−0.436	0.327	0.355	0.68	0.03	0.09	9.2	−1.4	1.5
R. Foot	0.115	0.217	0.074	0.19	0.03	−0.23	−2.1	1.0	−0.8
L. Foot	−0.579	0.316	0.097	0.32	0.06	0.26	6.0	0.7	3.3

Note. The XYZ positions refer to the global coordinate system defined in Figure 3.4. The velocities (in m/s) are v_X, v_Y, and v_Z, while a_X, a_Y, and a_Z are the accelerations (in m/s^2).

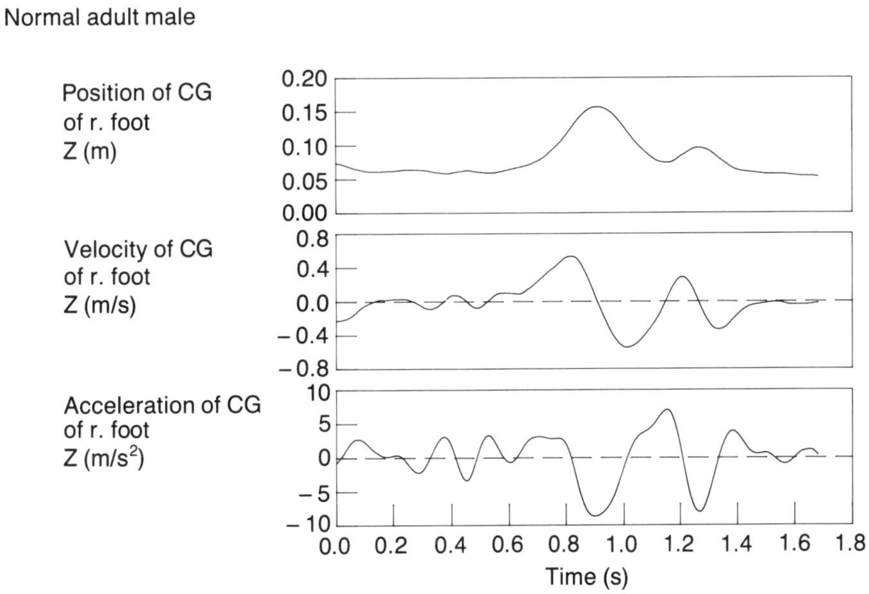

Figure 3.12 The displacement, velocity, and acceleration of the right foot's center of gravity in the Z direction as a function of time.

point), the acceleration must be zero. The position curve has turning points at about 0.91, 1.15, and 1.28 s; at these same points the velocity is zero. The velocity curve has turning points at about 0.83, 1.02, 1.21, and 1.33 s, and at these points, the acceleration is zero.

Methodological Notes

As a final comment, you should be aware that the digital filter algorithm has some endpoint problems (Vaughan, 1982). This means that the algorithm has a tendency to oversmooth the first few and last few data frames, which can result in erroneous velocity and acceleration data. One way of overcoming this is to sample extra frames of data on either side of the period of interest and then ignore these extra frames after the data have been smoothed. Another approach would be to use a different smoothing algorithm, such as a least squares quintic spline. The latter was considered for *GaitLab*, but the increased processing time and memory requirements mitigated against it (Vaughan, 1982).

Angular Kinematics

In this section, you will learn about two different ways to express the angular orientation of the segments in 3-D space. First, we will show you how one segment is oriented relative to another—the anatomical joint angles. Second, we will define how one segment is oriented relative to the fixed global reference frame—the segment Euler angles, named after the 18th-century Swiss mathematician. The anatomical joint angles are important because the ranges of movement are of interest to clinicians (e.g., hip abduction and adduction, knee flexion and extension, ankle inversion and eversion). The segment Euler angles are important because they are needed to define the angular velocities and angular accelerations of the segments. These latter two angular kinematic parameters are used in the equations of motion (see Figure 1.5) to calculate the joint moments.

Definition of Anatomical Joint Angles

There has been some debate as to the most appropriate method of defining joint angles so that they make sense from a traditional, anatomical point of view. We have decided that the most sensible method has been proposed by Chao (1980) and Grood and Suntay (1983). Consider the segment reference frames that were defined in Figure 3.10. Each joint has a reference frame in the proximal and distal segments (for the hip joint, this is the pelvis and thigh; for the knee joint, the thigh and calf; for the ankle joint, the calf and foot). Joint angles are defined as a rotation of the distal segment relative to the proximal segment. The rotations may be defined, in general, as follows:

- Flexion and extension (plus dorsiflexion and plantar flexion) take place about the mediolateral axis of the proximal segment (i.e., the z axes in Figure 3.10).
- Internal and external rotation take place about the longitudinal axis of the distal segment (i.e., the x axes in Figure 3.10).
- Abduction and adduction take place about a floating axis that is at right angles to both the flexion/extension and internal/external rotation axes.

These angle definitions can be a little more easily understood by referring to Figure 3.13, which illustrates the left knee.

The first three columns of Table 3.7 show the anatomical joint angles for $t = 0.00$ s for the Man.ANG file in *GaitLab*, whereas Figure 3.14 concentrates on the right hip joint. The mathematical details describing how the rotation axes of Figure 3.13 can be used to generate the data in Figure 3.14 and Table 3.7 are quite complex and are included in Appendix B.

Definition of Segment Euler Angles

Segment Euler angles play an important role in calculating segment angular velocities and accelerations. Earlier in this chapter, we discussed the need to have at least three noncolinear points on a segment to describe its position in 3-D space uniquely. Each segment (or free body) in 3-D space has six degrees of freedom: Six independent coordinates are necessary to define the segment's position. Three of these coordinates are the X, Y, and Z positions of the segment's center of gravity, which are illustrated for the right calf in Figure 3.15. The other three coordinates are the Euler angles and can be defined with the aid of Figure 3.16. (Your ability to think in 3-D will now be

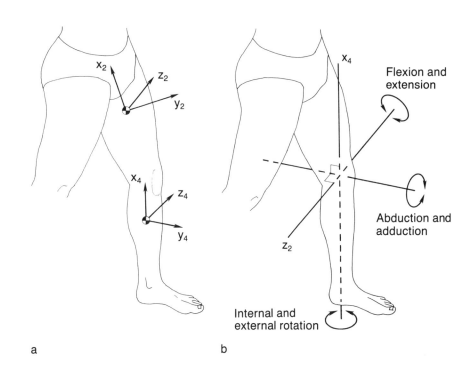

Figure 3.13 The axis of rotation for the left knee. The proximal and distal segment reference frames are shown on the left, while the axes are highlighted on the right. There are three separate ranges of motion: Flexion and extension take place about the mediolateral axis of the left thigh (z_2); internal and external rotation take place about the longitudinal axis of the left calf (x_4); and abduction and adduction take place about an axis that is perpendicular to both z_2 and x_4. Note that these three axes do *not* form a right-handed orthogonal triad, because z_2 and x_4 are not necessarily at right angles to one another.

tested to the limit!) First, we move the CG of the calf so that it coincides with the origin of the global reference system. We do this because the position of the calf's CG has already been taken care of as seen in Figure 3.15. Second, we define the *line of nodes*, which is a line at right angles to both the global reference axis Z and the calf axis $z_{R.Calf}$. Third, we rotate the xyz system of the calf from the XYZ (global) system to its actual position. The following three rotations are performed in order:

1. ϕ about the Z axis
2. θ about the line of nodes
3. ψ about the $z_{R.Calf}$ axis

These three rotations are the Euler angles and are illustrated in Figure 3.16. The particular convention that we have chosen to use—one based on the axes of rotation and their order (Z/line of nodes/z)—is that adopted by Synge and Griffith (1959) and Goldstein (1965) in their well-known texts on classical mechanics.

Angular Velocity and Acceleration

When the Euler angles have been calculated for each segment and at each instant in time, the angular velocities and accelerations may be derived.

$$\text{Angular velocity} = \text{Function [Euler angles and their first derivatives]} \quad (3.22)$$

$$\text{Angular acceleration} = \text{Function [Euler angles and their first and second derivatives]} \quad (3.23)$$

Table 3.7 Angular Kinematics (Anatomical Joint Angles, Segment Angular Velocities, and Accelerations) at Time = 0.00 s for a Normal Male

Joint/Segment	Flx/Ext	Abd/Add	Int/Ext	Omega$_x$	Omega$_y$	Omega$_z$	Omdot$_x$	Omdot$_y$	Omdot$_z$
R. Hip/Thigh	18.9	−5.0	−3.7	0.15	0.11	−0.08	9.3	6.5	−0.3
L. Hip/Thigh	−21.8	2.1	−13.3	0.61	−0.02	0.48	3.2	15.5	5.1
R. Knee/Calf	10.4	4.1	−1.7	−0.15	0.95	0.88	5.2	82.8	12.7
L. Knee/Calf	18.9	5.1	9.0	−0.19	0.60	−1.25	21.0	−9.4	−20.3
R. Ankle/Foot	−7.6	−6.8	−16.8	0.03	0.40	2.57	−0.7	3.1	−1.9
L. Ankle/Foot	−9.7	−8.3	−19.0	−1.80	−1.18	2.16	7.3	−16.6	38.4

Note. The first three columns are the anatomical joint angles. Refer to the text and Figure 3.13 for their definition. The middle three columns, the segment angular velocities, and the last three columns, the segment angular accelerations, are kinematic quantities derived from the segment Euler angles (refer to text and Figure 3.16) and are relative to the segment reference frames (Figure 3.10).

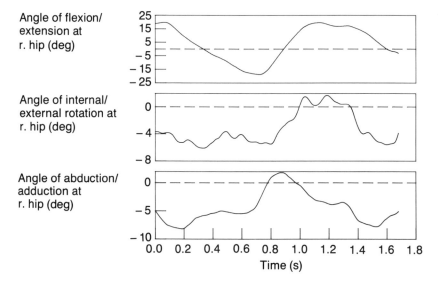

Figure 3.14 The orientation of the right thigh relative to the pelvis, that is, the right hip joint angles, plotted as a function of time. Top is flexion (positive) and extension (negative), middle is internal (+ve) and external rotation (−ve), and bottom is abduction (+ve) and adduction (−ve).

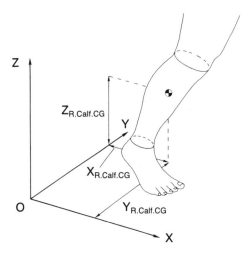

Figure 3.15 The three linear degrees of freedom ($X_{R.Calf}$, $Y_{R.Calf}$, and $Z_{R.Calf}$) defining the position of the right calf's center of gravity in terms of the global reference system XYZ.

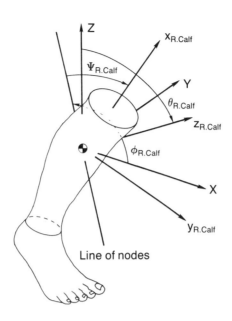

Figure 3.16 The three angular degrees of freedom (or Euler angles $\phi_{R.Calf}$, $\theta_{R.Calf}$, and $\Psi_{R.Calf}$) defining the orientation of the right calf's reference axes ($x_{R.Calf}$, $y_{R.Calf}$, and $z_{R.Calf}$) relative to the global reference system XYZ. Note that the calf's CG has been moved to coincide with the origin of XYZ. The three Euler angle rotations take place in the following order: (a) $\phi_{R.Calf}$ about the Z axis; (b) $\theta_{R.Calf}$ about the line of nodes; and (c) $\Psi_{R.Calf}$ about the $z_{R.Calf}$ axis. The line of nodes is perpendicular to both Z and $z_{R.Calf}$.

These equations are complex and lengthy, but have been included for the sake of completeness in Appendix B. There are a number of points that you need to be aware of, however.

1. The symbol for angular velocity is the Greek letter ω (pronounced "omega").
2. The angular velocity has three components and is expressed relative to a body-based coordinate system ($x_{R.Calf}$, $y_{R.Calf}$, $z_{R.Calf}$ in the case of Figure 3.16) rather than the global reference frame XYZ.
3. The angular acceleration is the first derivative of angular velocity, so we use the symbol $\dot{\omega}$ (pronounced "omega dot").
4. The units for angular velocity and acceleration are radians/second and radians/second^2 respectively (where radian is an angular measurement and is a dimensionless ratio).

Table 3.7 has an example of segmental angular velocities and acceleration at $t = 0.00$ s. Figure 3.17 is a plot of angular velocities and accelerations for the right calf about its own z axis.

Dynamics of Joints

So far in this chapter we have dealt only with kinematic quantities, such as displacement, velocity, and acceleration. As stressed in chapter 1, the kinematics are the effects, but we are also interested in studying the dynamics, the causes of motion. We now are ready to integrate body segment parameters, linear kinematics, centers of gravity, angular kinematics, and ground reaction forces in the equations of motion (Figures 1.4 and 1.5) to yield the resultant joint forces and moments.

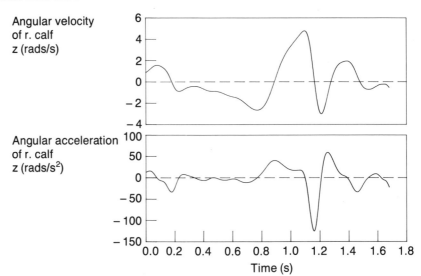

Figure 3.17 Angular velocity (top) and angular acceleration (bottom) of the right calf relative to its z axis, plotted as a function of time.

Measurement of Ground Reaction Forces

To perform 3-D gait analysis, we must have a force plate that provides six pieces of information:

1. Force in X direction, F_X
2. Force in Y direction, F_Y
3. Force in Z direction, F_Z
4. Position of resultant force in X direction, DX
5. Position of resultant force in Y direction, DY
6. Torque about Z axis, T_Z

Figure 3.4 shows the position of two force plates used in tandem where the origin of the global reference system XYZ is placed at the corner of Plate 1. Figure 3.18a depicts a foot on the plate, whereas Figure 3.18b shows the previously listed six pieces of information associated with the contact between the foot and plate.

Table 3.8 shows the ground reaction force information for both plates between 0.68 and 0.72 s for the Man.FPL file in *GaitLab*. Note that at 0.68 s, all the values for Plate 2 are zero. This simply means that the subject's left foot has not yet made contact with this plate. Note too that the DX value for Plate 2 is about 0.47 m greater than that for Plate 1. This simply reflects the fact that Plate 2 is that much further from the origin of the global reference frame (refer to Figure 3.4). Note also that F_Z in Plate 1 is decreasing (with increasing time) as the subject's right foot approaches toe-off. The ground reaction force data may be plotted as a function of time as seen in Figure 3.19. There are a few observations that need to be made when studying the F_X, F_Y, and F_Z curves in this figure. The range (in newtons) for the vertical force F_Z is almost three times that of the fore-aft force F_X. The mediolateral force F_Y has a range of less than one tenth of F_Z. The subject's weight is a little over 600N, so F_Z, which has the characteristic double hump, exceeds body weight at two different times during the stance phase.

As can be seen from all three curves, the stance phase lasts a little over 0.75 s. It should be pointed out that the male subject in this example was walking slightly slower than his normal pace (stance time is normally a little over 0.6 s). The fore-aft force F_X is negative (i.e., acting in a backward or aft

ANTHROPOMETRY, DISPLACEMENTS, & GROUND REACTION FORCES

Figure 3.18 The force plate used to measure the reaction forces of the ground acting on a subject's foot: (a) view of foot and plate showing XYZ global reference frame; (b) the resultant force $\mathbf{F}_R$ of the plate on the foot has three orthogonal components—F_X, F_Y, and F_Z. The position of this resultant force is specified by the coordinates DX and DY, and T_Z is the torque applied to the foot about the vertical Z axis.

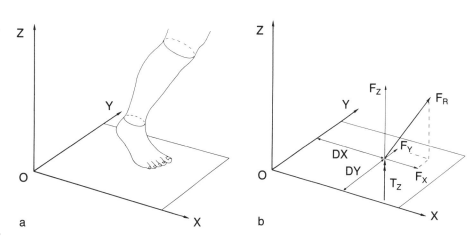

Table 3.8 The Three-Dimensional Force Plate Information, as Defined in Figure 3.18b, Showing the Data for Plates 1 and 2 Between 0.68 and 0.72 s

Time = 0.68 s Number	F_X (N)	F_Y (N)	F_Z (N)	DX (m)	DY (m)	T_Z (Nm)
Plate 1	108	−1	645	0.196	0.204	2.774
Plate 2	0	0	0	0.000	0.000	0.000
Time = 0.70 s Number	F_X (N)	F_Y (N)	F_Z (N)	DX (m)	DY (m)	T_Z (Nm)
Plate 1	117	−4	617	0.198	0.205	2.546
Plate 2	−15	30	271	0.672	0.272	0.609
Time = 0.72 s Number	F_X (N)	F_Y (N)	F_Z (N)	DX (m)	DY (m)	T_Z (Nm)
Plate 1	120	−2	538	0.199	0.206	1.417
Plate 2	−79	30	352	0.677	0.274	0.159

direction) for the first half of the cycle. Thereafter it becomes positive as the subject pushes off, driving backward on the plate, and, in accordance with Newton's third law of motion, experiencing a forward force. The mediolateral force F_Y is negative for the first 0.05 s, indicating that just after heel strike, the foot was acting on the plate in the positive Y direction (refer to Figure 3.18, and remember that all these forces—F_X, F_Y, F_Z—reflect the force of the plate on the subject's foot and not vice versa). During most of the stance phase F_Y is positive, which means that the plate is pushing inward (or in a medial direction) on the subject's foot.

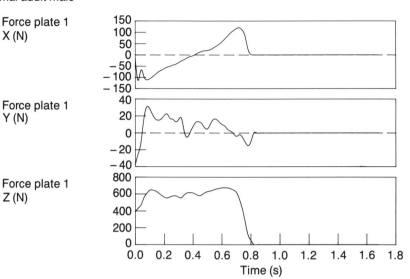

Figure 3.19 The fore-aft (F_X), mediolateral (F_Y), and vertical (F_Z) forces, in newtons (N), acting on the subject's right foot, as a function of time. For X, a forward force is +ve, backward is −ve; for Y, to the left is +ve, to the right is −ve; and for Z, up is +ve, down is −ve.

Calculation of Joint Forces and Moments

Now that we have the ground reaction force data, we can go on to calculate the resultant forces and moments acting at the joints of the subject's lower extremities—ankles, knees, and hips. The mathematics to accomplish this is not trivial (Greenwood, 1965), but we provide a simple overview of some of the key steps here in this chapter, leaving the details for Appendix B. First, you need to understand the concept of a Free Body Diagram (FBD). An FBD is a diagram in which the segment or body of interest (say, the right foot) is removed (or set free) from its environment and the external forces and moments acting on it are drawn in. Figure 3.20 shows an FBD for the right foot.

The next step is to apply Newton's second law of motion to each segment with the aid of the FBD. This law has two basic forms: linear and angular.

Linear: The summation of external forces acting on a segment (or free body) is equal to the rate of change of the linear momentum of the segment.

If we consider that

$$\text{Linear momentum} = (\text{Mass of segment}) \times (\text{Velocity of segment CG}) \quad (3.24)$$

and that the mass of the segment remains constant over time, the first derivative of Equation 3.24 may be written as

$$\text{Change in linear momentum} = (\text{Mass of segment}) \times (\text{Acceleration of segment CG}) \quad (3.25)$$

Using the definition for the linear form of Newton's second law of motion and Equation 3.25 we get

$$\Sigma \text{ Forces} = (\text{Mass of segment}) \times (\text{Acceleration of segment CG}) \quad (3.26)$$

Figure 3.20 Free Body Diagram for the right foot during pushoff. The *external* forces acting on the foot are its weight W_F, the resultant ground reaction F_R, and the force of the calf on the foot at the ankle joint $F_{R.Ankle}$. The external moments acting on the foot are the ground reaction torque about the Z axis T_Z, and the moment of the calf on the foot at the ankle joint $M_{R.Ankle}$. Note that the torque and moment are indicated as vectors with double arrow points. Compare this FBD for the right foot with Figure 3.18 and make sure you understand the similarities and differences.

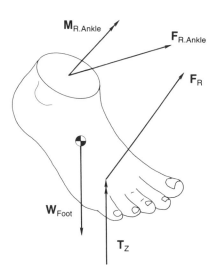

If we apply Equation 3.26 to the FBD in Figure 3.20, then the resulting equation is

$$F_{R.Ankle} + W_{Foot} + F_R = mass_{R.Foot} a_{R.Foot.CG} \qquad (3.27)$$

The only unknown in this equation is $F_{R.Ankle}$. The weight of the foot W_{Foot} is simply its mass ($mass_{R.Foot}$, obtained from Table 3.3) times the acceleration due to gravity g; the ground reaction force F_R is obtained from the force plate (Table 3.8 and Figure 3.19); the mass of the right foot $mass_{R.Foot}$ comes from Table 3.3; and the acceleration of the foot's center of gravity $a_{R.Foot.CG}$ was derived earlier (Table 3.6 and Figure 3.12). All this means that we now have sufficient information to solve for the resultant force at the right ankle joint $F_{R.Ankle}$. It is important to realize that this force is not simply the force of the tibia acting on the talar dome. Rather, it is the resultant of *all* the forces acting across the ankle, including bone, ligament, and muscular forces. To solve for all these individual forces uniquely is not possible because we do not have enough equations (Vaughan, Hay, & Andrews, 1982).

The angular form of Newton's second law of motion may be stated as follows:

Angular: The summation of external moments acting on a segment (or free body) about its center of gravity is equal to the rate of change of the angular momentum of the segment.

Expressed mathematically, and applied to the right foot of Figure 3.20, this law takes the form

$$\Sigma \text{ Moments} = \text{Rate of change of angular momentum} \qquad (3.28)$$
$$\text{of the segment}$$

We take moments about the segment's CG, so the left side of Equation 3.28 is

$$\Sigma \text{ Moments} = \mathbf{M}_{\text{R.Ankle}} + \mathbf{T}_Z + \\ \text{Moment due to } \mathbf{F}_{\text{R.Ankle}} + \\ \text{Moment due to } \mathbf{F}_R \quad (3.29)$$

The right side of Equation 3.28 has the form

$$\text{Rate of change of angular momentum} = \quad (3.30) \\ \text{Function [moments of inertia, angular velocity, and acceleration]}$$

The form of this function is well established (Synge & Griffith, 1959) and is presented in full in Appendix B. Because we know the principal centroidal moments of inertia (Table 3.3) and the segmental angular velocities and accelerations (Table 3.7), we can calculate the rate of change of angular momentum. The only unknown in the right side of Equation 3.29 is $\mathbf{M}_{\text{R.Ankle}}$. The ground reaction torque $\mathbf{T}_Z$ is known from the force plate (see Table 3.8, which contains data for the Man.FPL file used in *GaitLab*), the resultant force at the right ankle $\mathbf{F}_{\text{R.Ankle}}$ is calculated in Equation 3.27, and the ground reaction force $\mathbf{F}_R$ is obtained from the force plate (Table 3.8). In addition we know that

$$\text{Moment due to force} = \text{Force} \times \text{Lever arm} \quad (3.31)$$

Because we took moments relative to the segment center of gravity in Equation 3.29, the lever arms for $\mathbf{F}_{\text{R.Ankle}}$ and $\mathbf{F}_R$ are relative to the right foot's center of gravity. We know where the foot's CG is from Table 3.6, the right ankle's position from Table 3.5, and the point of application of the ground reaction force from Table 3.8. With this information, we calculate the lever arms and hence the moments due to the proximal force at the ankle and the distal force from the ground.

Using Equations 3.26 to 3.31, we can calculate $\mathbf{M}_{\text{R.Ankle}}$; $\mathbf{F}_{\text{R.Ankle}}$ was calculated in Equation 3.27. Now, by Newton's third law of motion (also known as the law of action and reaction), if we know the force and moment exerted by the calf on the foot at the ankle, then the force and moment exerted by the foot on the calf at the ankle has the same magnitude and opposite direction. We can then apply Newton's second law (Equations 3.26 and 3.28) to the calf to solve for the resultant force and moment at the knee joint. We repeat the process for the thigh to find the force and moment at the hip joint. Just as this procedure has been applied to the right side, it can be applied to the left side, providing the foot is either airborne (in which case the ground reaction force data are zero) or in contact with a force plate (and the ground reaction force data can be measured).

Expression of Joint Forces and Moments

The resultant joint forces and moments are three-dimensional vectors. This means that they can be expressed in terms of their components. One way of doing this is to use the global reference frame XYZ as the basis for the components. The drawback of this approach, however, is that it can be difficult to relate these laboratory-based components to human subjects, particularly those who walk at an angle to the X and Y axes instead of walking parallel to the X axis as illustrated in Figure 3.4. We believe a more sensible approach is to express the forces and moments in terms of body-based coordinate systems that have some anatomical significance. We have chosen the same axes used to define anatomical joint angles. These are as follows:

Forces

- A mediolateral force takes place along the mediolateral axis of the proximal segment.
- A proximal/distal force takes place along the longitudinal axis of the distal segment.
- An anterior/posterior force takes place along a floating axis that is perpendicular to the mediolateral and longitudinal axes.

Moments

- A flexion/extension moment takes place about the mediolateral axis of the proximal segment.
- An internal/external rotation moment takes place about the longitudinal axis of the distal segment.
- An abduction/adduction moment takes place about a floating axis that is perpendicular to the mediolateral and longitudinal axes (see Figure 3.21).

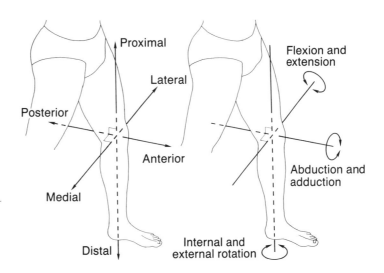

Figure 3.21 The reference axes of the left knee for expressing the components of (a) the resultant force at the knee; and (b) the resultant moment at the knee. Refer to the text for further discussion of these axes.

Table 3.9 shows the resultant forces and moments expressed in terms of these components (the table contains data for the Man.DYN file used in *GaitLab*), and Figure 3.22 concentrates on the flexion/extension moments at the right hip, knee, and ankle joints. These curves compare favorably with other data in the literature (Andriacchi & Strickland, 1985; Apkarian et al., 1989; Gainey et al., 1989; Winter, 1987). In Figure 3.23 the three components of the resultant moment at the right knee joint have been plotted. You will notice that the ranges (in newton • meters, n • m) for the flexion/extension and abduction/adduction moments are of the same order of magnitude. This is an excellent example of the potential danger in assuming that gait is purely a two-dimensional activity, and therefore casts some doubt on concepts such as the support moment proposed by Winter (1987). Note that in both Figures 3.22 and 3.23, the time axis runs from 0.00 to 1.40 s (i.e., from right heel

Table 3.9 The Three-Dimensional Resultant Joint Forces (F) and Moments (M) Acting at the Six Major Joints of the Lower Extremity at Time = 0.30 s

Joint	F_{MedLat}	F_{AntPos}	F_{PrxDis}	M_{FlxExt}	M_{AbdAdd}	M_{IntExt}
R. Hip	−10	−34	−674	6.1	49.5	3.0
L. Hip	6	−38	−111	−2.1	−0.6	−0.4
R. Knee	−76	85	−599	−19.5	15.3	−0.7
L. Knee	1	33	−32	−7.8	−0.9	−0.2
R. Ankle	−45	−568	57	17.9	1.6	−6.4
L. Ankle	1	−6	−9	0.0	0.2	0.1

Note. The forces are in newtons, N; the moments are in newton-meters, N•m.

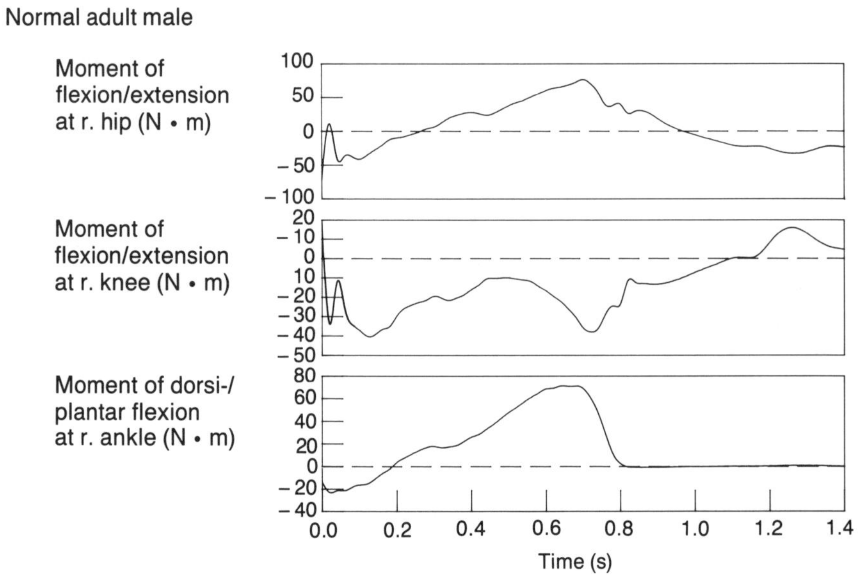

Figure 3.22 The flexion/extension moments (in newton-meters, N•m) at the right hip, knee, and ankle joints plotted as a function of time.

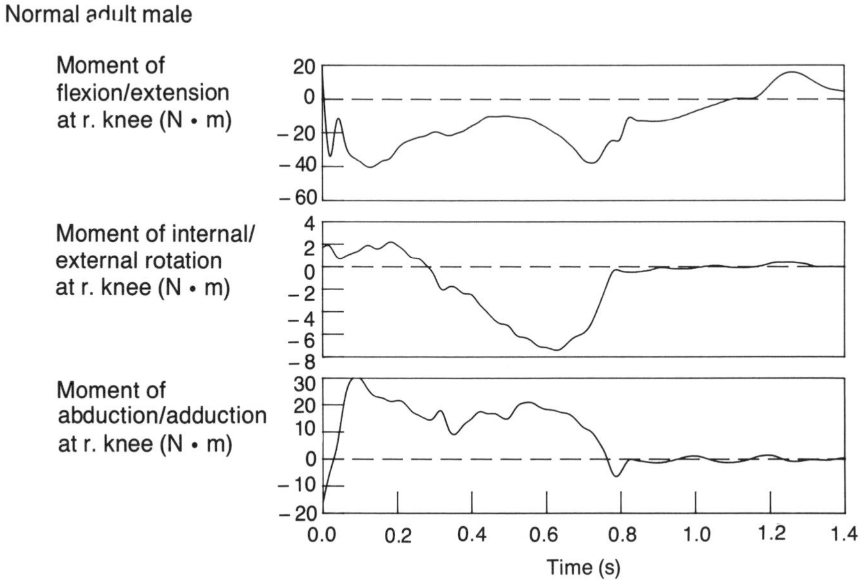

Figure 3.23 The flexion/extension, internal/external rotation, and abduction/adduction moments (in newton-meters, N•m) at the right knee joint plotted as a function of time.

strike to the next right heel strike). We do not have force plate information for the second right foot contact (although we do have kinematic data up to 1.68 s), so it is inappropriate to calculate joint forces and moments for the final 0.28 s.

Summary

We have finally reached the furthest point up the movement chain—the joint forces and moments—in our efforts to determine the causes of the observed movements. We cannot measure the tension in the muscles (see Figure 1.4) because there are not enough equations (such as 3.26 and 3.28) to calculate the large number of unknown muscle tensions. This state of indeterminacy has been solved by some researchers using mathematical optimization techniques (Crowninshield & Brand, 1981; Davy & Audu, 1987), but their predictions of individual muscle tensions have been only partially validated using electromyography. In chapter 4 you will learn some of the fundamentals of electromyography—particularly their applications to human gait.

CHAPTER 4

Muscle Actions Revealed Through Electromyography

To record shape has been far easier than to understand it.
C.S. Sherrington, 1940

"The time has come to take EMG seriously." This quote from Hof (1988) suggests two things: First, there are investigators who in the past have not taken EMG seriously; and second, there are good reasons why this should no longer be the case. We do not intend to suggest that EMG is the ultimate tool for understanding human gait, but we hope that you will find there are some definite uses for the technique in the field of locomotion studies.

Much of the confusion surrounding EMG analysis stems from an inadequate understanding of what is being measured and how the signal is processed, so we discuss some basic methodological issues first. These include basic electrochemistry regarding the operation of electrodes, selection of sampling frequencies, and signal processing methods. Next, we review the phasic activity of the major muscle groups involved in human gait. Finally, we study how these muscles interact with one another and reveal some basic patterns using a statistical approach.

Back to Basics

With the prospect of gaining some insight into the neuromuscular system, you may be tempted to rush in and apply any conveniently available electrode to some suitably prominent muscle belly, in the belief that anything can be made to work if you stick with it long enough. Rather than pursuing such an impetuous approach, we believe that the necessary attention should first be paid to some basic electrochemical principles.

Electrodes

To begin with, you need to have some idea of what an electrode is. Basically, it is a transducer: a device that converts one form of energy into another, in this case, ionic flow into electron flow (Warner, 1972). The term *electrode potential* has been defined as the difference between the potential inside the metal electrode and the potential at the bulk of the solution (Fried, 1973). This implies that the metal electrode cannot, by itself, be responsible for the electrode potential. Thus to prevent confusion, it is better to refer to *half-cell potentials*, which suggests that it is not just the electrode that is important, but the solution as well. Figure 4.1 shows a simple half-cell.

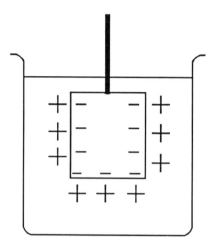

Figure 4.1 A simple half-cell.

In the half-cell shown in Figure 4.1, the metal ions tend to go into solution, and the electrolyte ions tend to combine with the metal. This results in a charge separation, which occurs in the region of the electrode and the electrolyte boundary. Because one can think of this boundary as being composed of two oppositely charged sheaths, it is commonly referred to as the *double layer*. In the ideal case, if two electrodes made of the same metal were in contact with a solution of their salts (e.g., zinc in a solution of zinc sulphate), the double layers would be identical, and there would be no net potential between the two electrodes. If the metals were different, such as copper and zinc electrodes, there would be a galvanic potential between the two (Figure 4.2). In the context of measuring EMG potentials, the electrodes must be identical, because you want the measured voltage difference to be attributable to half-cell changes due only to ionic flow within the electrolyte—not to galvanic potentials.

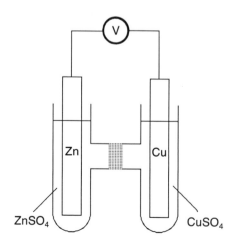

Figure 4.2 The potential (V) between two electrodes, one made of zinc/zinc sulphate and the other of copper/copper sulphate.

The concepts of half-cell potential and the double layer are extremely important for anyone working in the field of electromyography, because artifacts in the measured voltages can result when the double layer is disturbed by movement of the electrode. Warner (1972) describes a motion artifact as a percentage disturbance of the double layer. Electrode motion is unavoidable during walking, and especially during running; but if the artifact can be reduced somehow, the final EMG data will contain fewer errors. The easiest way to achieve minimal motion artifacts is by choosing suitable electrodes. An electrode with a half-cell potential of 1.6 V would have twice the artifact compared to an electrode with a half-cell potential of 0.8 V for an equal disturbance in the double layer (Warner, 1972). Typical half-cell potentials appear in Table 4.1.

Table 4.1 Typical Half-Cell Potentials for Different Metals That Could Be Used for EMG Electrodes

Electrode	Potential (V)
Zinc	−0.760
Tin	−0.140
Lead	−0.126
Hydrogen	0.000
Silver (Ag)	0.799
Platinum	1.200
Gold	1.420
Ag/AgCl	0.223

Note. Adapted from Rieger (1987).

It is apparent from Table 4.1 that pure silver has a half-cell potential of more than three times that of an electrode made of the same material but with a coating of silver chloride deposited on the surface. Pure silver is a poor electrode for EMG studies, because silver ions do not exist in the body nor in commonly used electrolytes. Therefore, the reaction at the boundary never reaches equilibrium. Large artifacts can also be caused by the migration of chloride ions from the body to the electrode surface. On the other hand, if the silver is coated with silver chloride, it is usual to obtain a stable half-cell potential of only 0.223 V. This, together with the fact that a stable equilibrium can be achieved between the chloride ions in the electrolyte and those in the body, allows EMG signals of moving subjects to be measured relatively accurately.

Warner (1972) describes a second possible source of artifact when measuring EMG signals from an exercising subject: the changes in the composition of the electrolyte resulting from the addition of sweat. Because perspiration is about 99.2% water, there will be a progressive shift in the baseline voltage as the amount of perspiration increases. A similar problem occurs in long-term studies in which the electrolyte is not completely sealed from the air and progressive drying occurs. In the latter case, crystals form and there are fewer free ions available to make up the half-cell (Warner, 1972).

To conclude this discussion of characteristics that make up a good electrode, it must be emphasized that the physical attachment of the electrode should be firm enough to restrict lateral motion and electrolyte drying, yet not so firm that it hinders the movement of the subject. The electrodes should be as similar as possible, with a low half-cell potential, such as found with the Ag/AgCl combination. If these are used, they should be cleaned gently, after each application, to keep the silver chloride coating intact.

Electrophysiology

Until now, we have tacitly assumed that there is some activity under the skin that is worth measuring. At this stage it is worthwhile to consider the underlying physiological processes that cause an electrical voltage to appear on a person's skin. Of necessity the discussion will be brief, but you are encouraged to refer to other excellent books on this topic (Loeb & Gans, 1986; Webster, 1976).

At the risk of oversimplifying the physiological processes that result in a nerve impulse being transmitted down a nerve axon, Figure 4.3a roughly illustrates the form of an axon potential. In this diagram a dipole (a small object with a positive and negative charge at each end) is moving along a volume conductor. A differential amplifier records the difference between the potentials at Points A and B on the conductor. Initially there is no difference between these points, so the output from the amplifier is zero. However, as the dipole approaches Point A, this point becomes negative with respect to B, and there will be some negative output. As soon as the dipole passes A, the situation changes. Not only does Point B start to register the approaching negative charge, but Point A begins to feel the effect of the positive charge at the other end of the dipole. The result is that when the quantity A-B is amplified, it is markedly positive. Finally, when Point B registers the positive end of the dipole, and Point A has returned to zero potential, the quantity A-B is slightly negative.

The resulting graph in Figure 4.3b is triphasic in appearance and bears some similarity to an action potential that actually traverses a nerve axon. It should be noted that a single axon leading to a muscle is responsible for the innervation of as few as 3 or as many as 2,000 individual muscle fibers (Winter, Rau, Kadefors, Broman, & DeLuca, 1980). A neuron and the muscle fibers that it innervates are referred to as a *motor unit*. Once an action potential reaches a muscle fiber, it propagates proximally and distally and is termed the *motor action potential*, or MAP. A *motor unit action potential* (MUAP) is a spatiotemporal summation of MAPs for an entire motor unit. Finally, an electromyographic signal (which is the topic of interest for the rest of this chapter) is the algebraic summation of many repetitive sequences of MUAPs

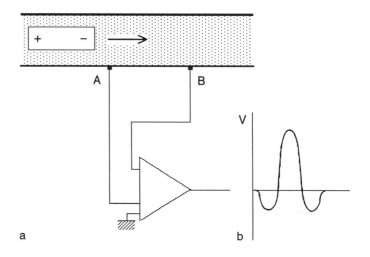

Figure 4.3 A dipole moving down a volume conductor (a) is analogous to the action potential that travels down a nerve axon. Refer to the text for an explanation of the way that the triphasic signal (b) is generated.

from all active motor units in the vicinity of the recording electrodes (Winter et al., 1980).

It is also important to note that the commands given to a muscle do not always cause the muscle to shorten. Cavanagh (1987) notes that one of the main functions of a muscle may be indeed to act as a brake, rather than to shorten actively. This is one reason why he prefers the term muscle *action* to muscle *contraction*. In this chapter we too will use the terminology he has suggested and use *eccentric* and *concentric action*, where appropriate.

Selection of Sampling Frequency

Once you have decided to measure myoelectric signals from a muscle and have applied a suitable electrode to the surface of the skin, the issue of sampling frequency needs to be addressed. The importance of this issue can be seen in Figure 4.4. If the actual signal has a cyclic period of 0.033 s (i.e., a frequency of 30 Hz) and you sample every 0.016 s (i.e., a frequency of 60 Hz), it is theoretically possible to record no signal at all. In passing, it is important to point out that sampling at 60 Hz is extremely inadvisable, because this is the frequency at which electricity is generated in many parts of the world. It is therefore very likely that any recording would be more a reflection of the electrical fields surrounding the equipment in the laboratory than of the physiological processes within the muscle of interest. To circumvent the problem of choosing a sampling frequency that is too low, the theorem of Shannon is often used as a rule of thumb: The sampling frequency should be at least twice the highest frequency of interest. According to standards prepared under the auspices of the International Society of Electrophysiological Kinesiology (ISEK), the range of signal frequencies for surface EMGs is from 1 to 3,000 Hz (Winter et al., 1980). This means that to prevent the problem of aliasing illustrated in Figure 4.4, a sampling frequency of at least 6,000 Hz may be required. In general, however, most of the power of the signal is in the range 50 to 150 Hz and certainly below 250 Hz (see Figure 4.5). For this reason a sampling frequency of 500 Hz would be more than adequate for surface EMG and should be within the capabilities of most data capturing systems presently in use.

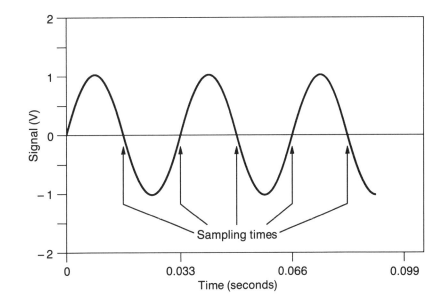

Figure 4.4 Although this signal has a frequency of 30 Hz and sampling is at 60 Hz, it is still possible to get no signal. This problem is known as "aliasing."

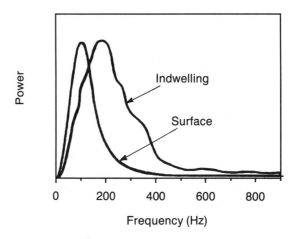

Figure 4.5 The power in an EMG signal plotted as a function of the frequency content. Notice that indwelling electrodes have a higher frequency content. *Note.* Based on data from Winter (1979).

Surface Versus Indwelling Electrodes

The electrode options available to someone wishing to collect EMG data are many and varied. You can use needle or wire electrodes (Basmajian & DeLuca, 1985) and there are other possibilities, such as monopolar or bipolar surface electrodes. For a comprehensive review of these and other electrode types, refer to the texts by Loeb and Gans (1986) and Geddes (1972). Some examples of surface electrodes are illustrated in Figure 4.6. The advantages of these electrodes are that they are simple to use, are noninvasive, and, if the skin surface is well prepared, will provide a good indication of the underlying muscle activity. (Preparing the skin surface requires that hair be shaved and that the dry, keratinized layer be removed by lightly abrading the skin with very fine sandpaper and then swabbing with alcohol-soaked cotton.) The disadvantage of surface electrodes, however, is that the activity of deep muscles cannot be monitored. Though in many applications in gait analysis this is not a problem, there are times when a deep muscle, such as tibialis posterior, may be suspected of some underlying pathology (such as spastic hemiplegia with a varus foot) and only deep, indwelling electrodes can be used.

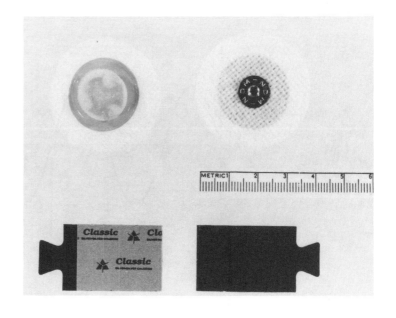

Figure 4.6 Some examples of commercially available surface electrodes used to capture EMG signals. These electrodes are used more often to capture ECG (electrocardiographic) signals, but they also work well for EMG.

When using bipolar surface electrodes, it is typical to have the signals preamplified and then sent to a main differential amplifier. A possible source of confusion here is that if the amplifier measures the difference between the two signals at its input stage, the need for a separate ground electrode is not that obvious. However, the way that most differential amplifiers are designed (see Figure 4.7), a separate ground electrode is needed to ensure that the input signals are referenced in such a way that they do not exceed the limits of the amplifier. A more detailed discussion on amplifiers may be found in Loeb and Gans (1986).

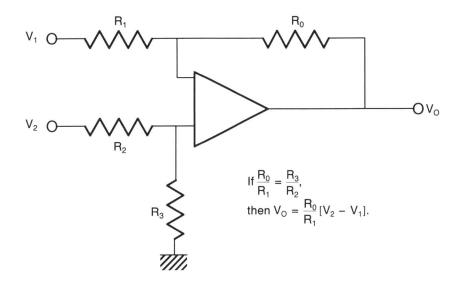

Figure 4.7 A simple differential amplifier. The signals V_1 and V_2 are the input from the electrodes placed on the muscle; the signal V_O is the output from the amplifier.

If $\dfrac{R_0}{R_1} = \dfrac{R_3}{R_2}$, then $V_O = \dfrac{R_0}{R_1}[V_2 - V_1]$.

As mentioned earlier, one method for reducing motion artifact is to select electrodes such as the Ag/AgCl variety, which have low half-cell potentials. Another method of getting rid of most of these spurious signals is to take advantage of the fact that motion artifact noise is at the low end of the frequency spectrum (as seen in chapter 2, most gait signals repeat about once per second, i.e., one hertz). By filtering out or removing any signals with a frequency less than 20 Hz, many of the problems caused by artifacts can be reduced. Also, as indicated previously, using a notch filter to eliminate 60 Hz background noise (sometimes referred to as "mains hum") would be advisable.

Signal Processing Methods

Some methods for processing EMG data have been recommended by ISEK (Winter et al., 1980). As seen in Figure 4.8, the raw EMG signal may be processed in a number of ways. These include

- full-wave rectification, in which the absolute value of the signal is taken;
- a linear envelope detector, which consists of full-wave rectification followed by a low-pass filter (i.e., all the high frequencies are removed);
- integration of the linear envelope (taking the area under the curve) over the period of interest; and
- a simple binary threshold detector, in which the muscle is designated to be either *off* or *on*.

Figure 4.8 Some of the most common methods for processing the raw EMG signal. Notice that the threshold detector to determine if a muscle is *on* or *off* must be set arbitrarily. *Note.* Adapted from Winter (1979).

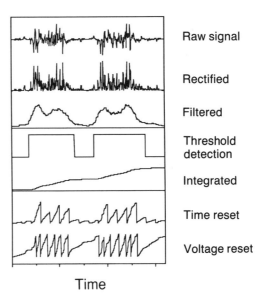

For the rest of this chapter, and in the software examples demonstrated in chapter 5 and *GaitLab*, we have chosen to represent EMG signals processed by the linear envelope method.

Phasic Behavior of Muscles

In Figure 2.11, you saw the EMG activity in a single muscle—a normal female's rectus femoris—plotted as a function of the gait cycle. One of the fascinating features of human gait, however, is that the central nervous system must control many muscles simultaneously. Figure 4.9 shows the normal EMG patterns for 28 of the most important muscles in the lower extremities plotted as functions of the gait cycle. When you consider that this graph is for one side of the body only and that there is another set of muscles on the other side which are half a cycle out of phase, you realize just how complex the human locomotor apparatus is!

The order of the muscles in Figure 4.9 has been chosen so that there appears to be a wave of muscle action that flows from left to right, that is, from heel strike through to the next heel strike. Muscles with similar phasic activity have been grouped together. This applies both to muscles with similar actions (such as tibialis anterior and extensor digitorum longus), as well as those with no immediately apparent connection (such as rectus femoris and gluteus maximus).

Inman et al. (1981) introduced a novel method for illustrating the actions of leg muscles during the gait cycle. Wooden models of the pelvis and lower limbs were constructed and arranged in an expanded and sequential series depicting a single stride. Based on photographs of these models, drawings were made, and muscle groups were superimposed on the drawing of each model at each position. Then the level of the muscle activity was indicated by color: red, highly active; pink, intermediate; and white, quiescent. We have adapted and extended their concept in this book. Figure 4.10 shows a gait cycle from two separate views, posterior and lateral; the events are indicated using the conventions described in Figures 2.5 and 2.7. The seven major muscle groups represented in Figure 4.10 are

1. gluteus maximus (posterior view);
2. gluteus medius (posterior and lateral views);

MUSCLE ACTIONS REVEALED THROUGH ELECTROMYOGRAPHY

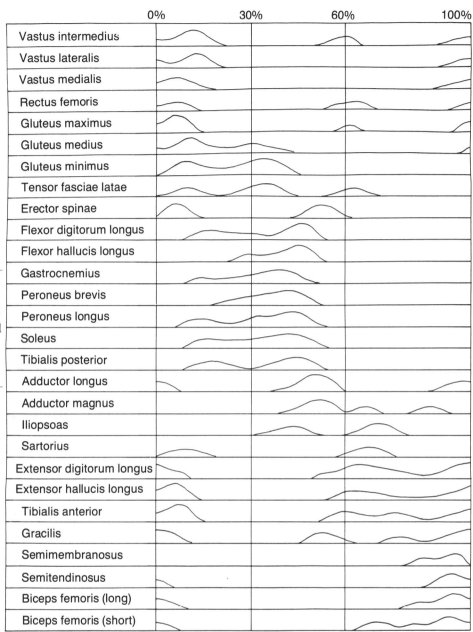

Figure 4.9 Normal EMG patterns for 28 of the major muscles in the lower extremities plotted as a function of the gait cycle. *Note.* Adapted from Bechtol (1975).

3. adductor magnus (posterior view);
4. quadriceps (lateral view);
5. hamstrings (posterior and lateral views);
6. triceps surae (posterior and lateral views); and
7. tibialis anterior (lateral view).

The shading indicates the degree of activity: black, most active; stippled, intermediate; and white, quiescent. The images in Figure 4.10 form part of the animation sequences that are the focus of Appendix A, where we will extend the concepts of Inman et al. (1981) and enable you to bring the human locomotor apparatus to life by fanning the pages of this book. In addition, these muscle activity sequences have been color-coded and animated in *GaitLab*.

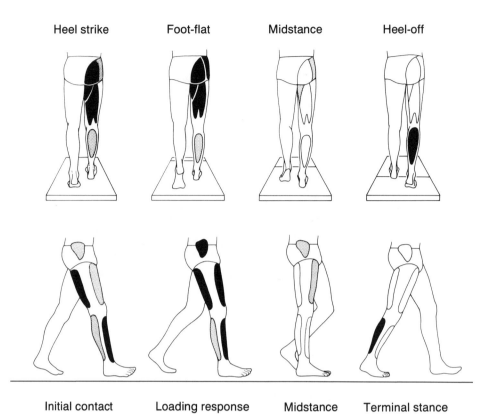
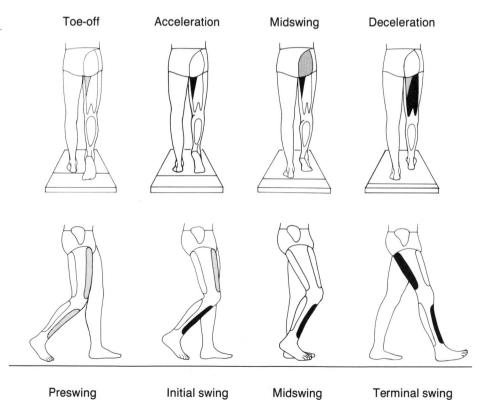

Figure 4.10 Posterior and lateral views of seven of the major muscles of the lower extremities, showing the activity of the muscles at key phases of the gait cycle.

A careful study of Figures 4.9 and 4.10 allow certain generalities to be made concerning the phasic behavior of the muscles. Most of the major muscle groups are active at or around both heel strike and toe-off (i.e., at the beginning and end of the stance and swing phases of the cycle). These are the periods of deceleration and acceleration of the legs, when body weight is transferred from one foot to the other. During midstance and midswing, most muscles (with the exception of gluteus medius and triceps surae during stance, and tibialis anterior during swing) are relatively quiescent. This is interesting because it is during these two periods (midstance and midswing) that the greatest observable movement takes place. During midstance, gluteus medius acts as a hip abductor to stabilize the pelvis as the contralateral leg swings through, while the triceps surae prevents excessive dorsiflexion of the ankle and then prepares to drive the person forward. During midswing, the tibialis anterior (as well as extensor digitorum longus and extensor hallucis longus) provides active dorsiflexion and thus prevents the toes from dragging on the ground. As a general rule, then, it appears that one of the principal actions of the muscles is to accelerate and decelerate the angular motions of the legs (Inman et al., 1981).

Relationship Between Different Muscles

Although the foregoing review considered the phasic activity of all the major muscles separately, some useful insights into the neuromuscular system can be obtained by studying the relationships that exist between different muscles. A question frequently of interest to those involved in gait analysis concerns the degree to which muscles are synergists or antagonists. There seems to be little dispute in the literature concerning the phasic behavior of the plantar flexors and dorsiflexors of the ankle during natural walking. Mann and Hagy (1980) found that when the one group was *on* the other was *off*, a finding endorsed by Procter and Paul (1982) and Inman et al. (1981), and seen in Figures 4.9 and 4.10. This pattern can be demonstrated by plotting the activity levels of one muscle as a function of the other, and Figure 4.11 illustrates such a graph. Note that when each muscle is highly active the other is almost quiescent.

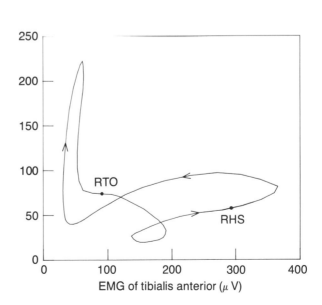

Figure 4.11 A phase diagram of the EMG of the triceps surae plotted as a function of the EMG of the tibialis anterior, both in microvolts (μV). Note the L shape of the curve, which indicates that these two muscles act in a reciprocating manner (when one is active, the other is quiescent and vice versa). RTO, right toe-off; RHS, right heel strike.

Figure 4.11 is based on data published by Winter (1987) and was generated in *GaitLab*. We recommend that you experiment with all the possible combinations of pairs of muscles (note that *GaitLab* has data for eight muscles, so there are 28 possible pairings of different muscles). Many different forms of relationships will be found. Figure 4.12 shows that gluteus medius and rectus femoris are relatively coactive. If their activity levels increased and decreased in perfect harmony, the curve would be a straight line passing through the origin and the top right-hand corner of the graph. If, in addition, the slope of the curve was unity, it would indicate that the magnitudes at any particular instant were the same. When there is a loop in the graph (as seen in Figure 4.13), you can deduce that there is some phase lag between the two individual EMG traces. Although a graph such as that shown in Figure 4.13 may be difficult to interpret, the effort of thinking about gluteus medius and hamstrings would most likely help you gain a better understanding of the relationship between these muscles.

Figure 4.12 A phase diagram of the EMG of the rectus femoris plotted as a function of the EMG of the gluteus medius. Note that these two muscles, although they have quite different roles, are almost perfectly in phase with one another. RTO, right toe-off; RHS, right heel strike.

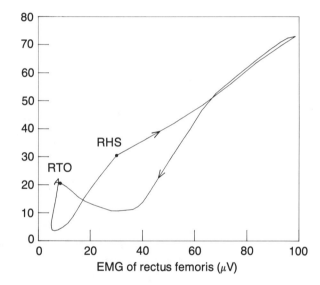

Figure 4.13 A phase diagram of the EMG of the hamstrings plotted as a function of the EMG of the gluteus medius. The large loop indicates that there is a considerable phase lag between these two muscles. RTO, right toe-off; RHS, right heel strike.

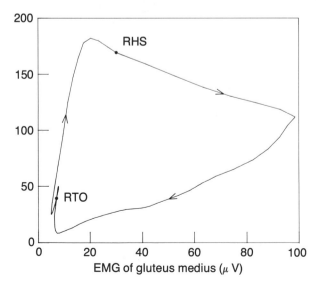

Multidimensional Scaling and Muscle Mapping

Although graphs such as Figures 4.11 to 4.13 allow the degree of coactivity to be easily visualized, the number of pairings becomes excessive when dealing with more muscles. For example, Winter (1987) listed the EMG activity for 16 muscles of the locomotor system, which would require 120 graphs. The geographer Gould (1985) used a novel way of depicting the relationships between many variables, called multidimensional scaling. (He used the technique to describe the "closeness" of academic journals—by noting how many times one journal cites another—and also the relationships between all the characters of *Romeo and Juliet*!) His statistical technique allows one to map the location of points on a set of axes such that the distance between the points gives some indication of how close those variables are to one another. In the present context, to find the best two-dimensional representation of 16 points would take an extremely long time, not to mention the tediousness and frustration that would be involved. Fortunately, Gould described a computerized method for optimizing the location of points that represent some complex relationship. This method involves an iterative procedure that "seeks" the best solution given some arbitrary starting configuration.

The degree of coactivity between the 16 muscles that Winter (1987) listed can be determined by following the procedure outlined in Figure 4.14. Figure 4.15 illustrates the resulting muscle map and shows the relationships between all 16 muscles in muscle space. The distance between any two muscles gives an indication of the similarity between those muscles.

From the map of muscle relationships in Figure 4.15, some interesting observations can be made. First, muscles with relatively low levels of EMG activity, such as gluteus maximus, sartorius, rectus femoris, and gluteus medius, are located toward the periphery. Conversely, muscles that are highly active, such as gastrocnemius or tibialis anterior, are much closer to the center. It is also interesting to note the various groupings of muscles. As the name *triceps surae* suggests, there are three calf muscles grouped together.

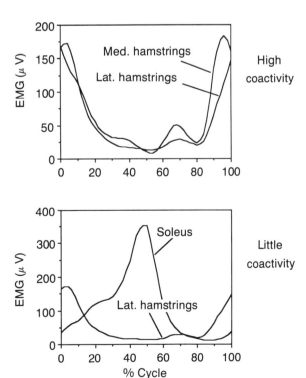

Figure 4.14 Prior to performing multidimensional scaling it is necessary to determine the coactivity between all pairs of muscles.

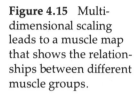

Figure 4.15 Multidimensional scaling leads to a muscle map that shows the relationships between different muscle groups.

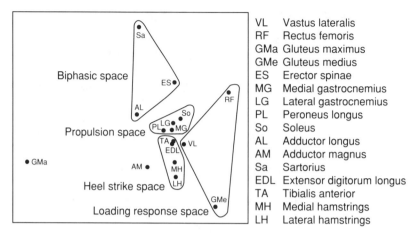

However, it is less obvious that tibialis anterior and extensor digitorum longus are associated with the hamstrings, although when you consider what happens at heel strike, this grouping does make sense: Just when the anterior calf muscles are doing their utmost to prevent the foot from slapping on the ground, the hamstrings are controlling and stabilizing the hip and knee joints. Further insight into the four main groups or spaces in Figure 4.15 will be provided a little later in this chapter when we introduce you to a factor analysis approach.

The position of peroneus longus in Figure 4.15 indicates that it is acting as an intermediary, bridging the activities of the pre- and posttibial muscles. This is not unexpected if one bears in mind what Winter (1987) has to say regarding the activity of this muscle: "The first burst appears to stabilize the ankle against foot inversion (possibly as a co-contraction to the tibialis anterior), and the remainder of the activity through to and including push-off is as a plantar flexor" (p. 54).

Factor Analysis

One of the goals of studying EMG is to discover the control processes involved in regulating muscular activity during gait. In other words, we want to uncover some of the traits of the homunculus, the "little man" responsible for coordinating our movement patterns (see Figure 1.1). Unfortunately, not much progress has been made in this regard, because neural signals cannot be measured directly and the complexity of the human nervous system far exceeds any electronic or computer device. The best we can do is to analyze EMG signals using either pattern recognition algorithms (Patla, 1985) or a branch of multivariate statistics commonly known as factor analysis (Kaiser, 1961).

Of relevance to the present chapter is the question, Are there some underlying trends that in combination can explain all the variations in a set of EMG traces? This question is important, because if it is found that there are fewer factors than there are muscles, the neural system may not need one program for each EMG signal. Thus, by decomposing the original data, we can ascertain how many independent patterns are needed to span the entire muscle space, and therefore how many EMG traces are needed to reconstruct *all* of the EMG patterns. Another advantage of this approach is that it permits us to know, for example, how similar gluteus medius is to vastus lateralis over a variety of instances in the gait cycle. The resulting correlation between EMG patterns and the underlying factors is called the *loading matrix*. If two EMG signals load highly on the same factor, then their patterns are similar.

A factor analysis, using the commercially available SAS package, was conducted on the data published by Winter (1987) for 16 leg muscles. The results of the analysis showed that only four main patterns emerged. These four patterns, or components, accounted for 91.5% of the variance. Whether to include a fifth pattern was debatable: It only accounted for an extra 5% of the variance in the EMG data. By including this fifth component, the total amount of variance accounted for would obviously have been 96.5%, but we decided that 91.5% was acceptable. The question to be addressed at this stage is, What do these four factors mean from a physiological point of view? To tackle this question, we need to look at the matrix of factor loadings shown in Table 4.2.

Table 4.2 The Analysis of 16 EMG Signals Using Multivariate Statistics Yields a Matrix of Loading Factors

Muscle	Factor 1	Factor 2	Factor 3	Factor 4
Gluteus medius	0.17	0.03	**0.98**	0.02
Gluteus maximus	0.57	−0.11	0.71	0.37
Medial hamstrings	**0.85**	−0.26	0.19	−0.07
Lateral hamstrings	**0.89**	−0.10	0.35	0.13
Erector spinae	0.29	0.19	0.35	**0.76**
Sartorius	0.09	−0.29	0.19	**0.85**
Rectus femoris	0.23	0.07	**0.87**	0.40
Vastus lateralis	0.31	−0.01	**0.89**	0.29
Adductor longus	0.15	−0.14	0.18	**0.93**
Adductor magnus	0.27	−0.76	0.23	0.05
Tibialis anterior	**0.86**	−0.24	0.17	0.36
Extensor digitorum longus	**0.83**	−0.15	0.28	0.38
Medial gastrocnemius	−0.26	**0.81**	0.10	−0.35
Lateral gastrocnemius	−0.11	**0.98**	−0.01	−0.03
Soleus	−0.40	**0.88**	0.02	−0.20
Peroneus longus	0.38	**0.77**	0.30	0.24

Factor Loadings—Patterns of Muscle Activity

Inspection of Table 4.2 offers some insight into how the EMG signals can be grouped. Both lateral and medial hamstrings as well as tibialis anterior and extensor digitorum longus load highly on the first factor. Thus, this factor can be thought of as describing the activity of those muscles whose peak EMG signal occurs within 5% of heel strike (i.e., 95% to 5% of the cycle). We have therefore called it the *heel strike factor*. The second factor describes the activity of the triceps surae and peroneus longus muscles, which are most active during mid- and terminal stance. This factor we called the *propulsion factor*. Muscles that load highly on the third factor are gluteus medius, rectus femoris, vastus lateralis, and possibly gluteus maximus. The similarity between these muscles is that they have peak activity levels at about 10% of the gait cycle, and, excepting the first, all have a slight, second burst just after toe-off. These muscles can be thought of as predominantly hip and thigh stabilizers, and this factor is therefore referred to as the *loading response factor*. Lastly, with regard to the fourth factor, it is apparent that sartorius, adductor longus, and possibly erector spinae are correlated. The similarity between these three muscles may be difficult to imagine, but by referring to Figure

4.9, it can be seen that these muscles all have two approximately equal peaks—just after heel strike and just after toe-off. This factor has therefore been called the *biphasic factor*. It is important to point out here that the heel strike factor is the most important, in the sense that it accounts for most of the variance, whereas the biphasic factor is the least important.

Further manipulation of the data (beyond what is shown in Table 4.2) yields the basic shape of the factor scores presented in Figure 4.16. Remember that by combining these four graphs, most of the trends in the 16 EMG traces can be described. It is interesting to compare these results with those of Patla (1985) who suggested that just four basic features were required to synthesize seven muscular activity patterns within various conditions of speed and stride length. In his analysis, the error in using only four features was 8.3%, which is similar to the error of 8.5% that our four factors have yielded. Wootten, Kadaba, and Cochran (1990a, 1990b) represented the linear envelope signals of 10 different muscles using a principal component analysis. They found that between 3 and 5 features (or components) were necessary to represent, and thereby reconstruct, the individual muscle patterns. Their purpose, however, was other than finding the relationship between different muscle groups (as we have explored in this chapter).

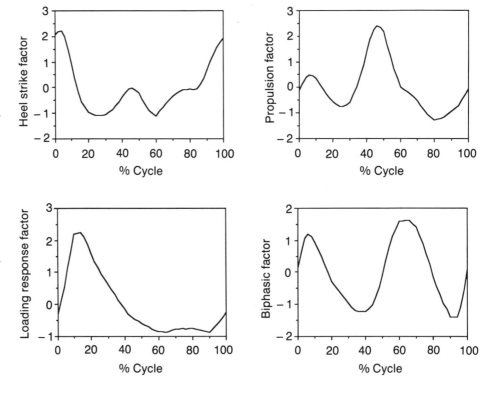

Figure 4.16 The four underlying factors (or trends) for the EMG profiles of 16 leg muscles based on multivariate statistics (see Table 4.2). Note that the units have been normalized, or *scaled*.

It is interesting to note how strongly the muscle map in Figure 4.15 ties in with the factor analysis in Table 4.2. The four factors in Table 4.2 have been represented as four spaces in Figure 4.15. If we choose a representative muscle from each of the four factors and plot its activity as a function of the gait cycle (see Figure 4.17), then the agreement with the factor score graphs in Figure 4.16 is remarkable. However, we must finish this discussion on the relationship between different muscles with a word of caution. Both the multidimensional scaling technique (Figures 4.14 and 4.15) and the factor analysis with multivariate statistics (Table 4.2 and Figure 4.16) were based

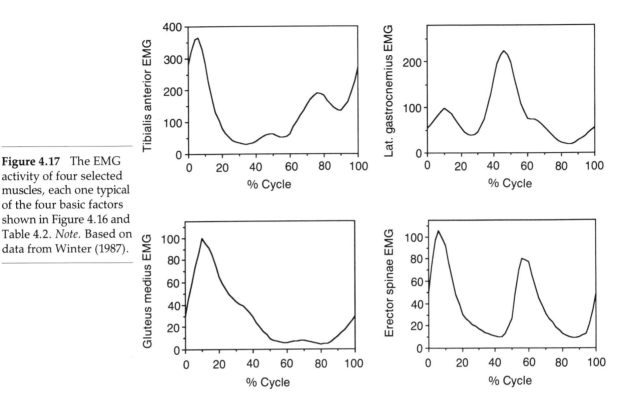

Figure 4.17 The EMG activity of four selected muscles, each one typical of the four basic factors shown in Figure 4.16 and Table 4.2. *Note.* Based on data from Winter (1987).

on *surface* EMG from Winter (1987). Surface EMG signals are notorious for the amount of crosstalk between muscles (Inman et al., 1981; Loeb & Gans, 1986). This means, for example, that a pair of electrodes placed over the soleus will almost certainly pick up the signal from nearby muscles, such as the two heads of the gastrocnemius, at the same time. The high degree of correlation between certain muscles may therefore be a result of the data capture procedure rather than a true manifestation of the underlying neuromuscular control system. One solution to this problem would be to gather fine-wire EMG data, but the likelihood of attracting a large pool of subjects would be rather slim. Despite the shortcoming of surface EMG, we believe that the foregoing discussion should give you a deeper appreciation of the control strategies that permit bipedal humans to walk in a smooth, coordinated fashion.

Summary

In this chapter on electromyography, we have taken you back to the basics—explaining some fundamental electrophysiology, the importance of understanding electrodes, and the issues of sampling frequency plus signal processing. In Figure 4.9 you saw the actions of 28 muscles and realized just how complex the human locomotor apparatus is. Finally, you learned how two statistical techniques—multidimensional scaling and factor analysis—can help reveal some basic patterns of interaction between muscles. In the next chapter, you will see how electromyography—and other gait analysis parameters—may be applied to the understanding of pathological gait.

CHAPTER 5

Clinical Gait Analysis— A Case Study

As we stated earlier, one of the purposes of this book is to introduce clinicians to the insights that gait analysis can offer when analyzing the locomotor status of patients. Our clinical example is a 23-year-old man, whom we will call SI. He was born by breech delivery, and his breathing was delayed. He had abnormal motor development and began physical therapy at the age of 13 months. By 20 months he was able to stand independently, and by 26 months he was walking. There was some disagreement among SI's physicians concerning his diagnosis. At age 15, the neurologist described a "steppage" gait, high arched feet, weakness in the calf muscles, and depression of ankle joint reflexes. Based on electromyographic data and nerve conduction velocity tests, the neurologist made a diagnosis of hypertrophic peripheral neuropathy of the type Charcot-Marie-Tooth. At the same age, SI's orthopaedic surgeon diagnosed him as having cerebral palsy with evidence of athetosis and hypotonia. He described SI's gait as being athetoid with instability of the ankle joints and an equinus foot on the left side. Although the surgeon believed that SI had cerebral palsy, he did not rule out hereditary hypertrophic polyneuropathy as an alternative diagnosis.

The orthopaedic surgeon elected to address the unstable ankles by performing an arthrodesis of the subtalar joints according to the method of Lambrinudi (1927). In this procedure, a wedge of bone is excised from the plantar aspect of the head and neck of the talus. The distal sharp margin of the body of the talus is inserted into a prepared trough in the navicular. The talus is thus locked in an equinus position at the ankle joint, whereas the rest of the foot maintains the desired degree of dorsiflexion. The procedure was done first on the left side and then on the right side (4 months later) when SI was 16 years old. Figure 5.1, a-h shows the pre- and postoperative X rays of his feet. Notice the clawing of the toes and the high arch before surgery. After surgery, the bones of the feet are in a more neutral alignment, and the staples used in the arthrodesis have provided the necessary stability.

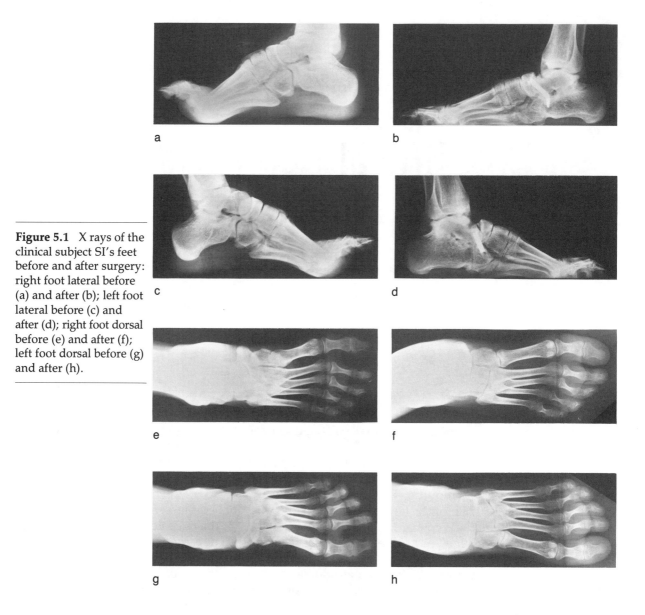

Figure 5.1 X rays of the clinical subject SI's feet before and after surgery: right foot lateral before (a) and after (b); left foot lateral before (c) and after (d); right foot dorsal before (e) and after (f); left foot dorsal before (g) and after (h).

Experimental Methods

We studied SI's gait when he was 23 years old and a very functional ambulator. His only difficulties were ascending and descending stairs (a result of the weakness in his calf muscles) and walking on uneven terrain (compromised by lack of mobility in his subtalar joints). We used a simple beam caliper and flexible tape to measure the necessary anthropometric data (see Tables 3.1 and 3.2). The electromyographic, kinematic, and force plate data were all gathered simultaneously. It should be noted that SI's EMG and force plate data were sampled at 1,000 Hz but are translated and reported in *GaitLab*'s Clinical subdirectory at 50 Hz. This was done to facilitate comparison with the kinematic data sampled at 50 Hz.

Kinematic Data

Fifteen spherical markers, 15 mm in diameter, were attached to SI's body with double-sided tape according to the marker configuration illustrated in Figure 3.10. A VICON system, consisting of strobed, infrared light and five cameras operating at 50 Hz, was used to capture the 3-D kinematic data (see Appendix C).

EMG Data

For the EMG, we shaved the hair from the area of interest and used surface Ag/AgCl pregelled electrodes with the Transkinetics telemetry system (see Appendix C). Only muscles on the right side were studied, and these included erector spinae, gluteus maximus, gluteus medius, lateral hamstrings, rectus femoris, adductor longus, tibialis anterior, and triceps surae. Details on the placement of the electrodes may be found in Winter (1987). The raw EMG data were fed through a cascade of three hardware filters: first-order high pass (20 Hz), third-order low pass (300 Hz), and sixth-order high pass (20 Hz) Butterworth filters. The analog signals were sampled at 1,000 Hz and then full-wave rectified and passed through a linear envelope detector in software (see Figure 4.8).

Force Plate Data

The ground reaction forces were measured with a pair of AMTI force plates arranged in tandem. Details on these devices, which monitor the six components of the ground reaction (refer to Figure 3.18), may be found in Appendix C. As mentioned, the EMG, kinematic, and force plate data were gathered simultaneously for three separate walking trials. Based on videotapes taken at the same time, we selected one representative trial for detailed analysis.

Results and Discussion

All the data that we collected on SI—the APM, EMG, KIN, and FPL files—are contained in *GaitLab*'s Clinical subdirectory. Therefore, all the data and figures that we present in this chapter are available to you. Although the following data and discussion are self-contained, you are encouraged to experiment with *GaitLab*. By exploring on your own you will develop a much better appreciation of the power of 3-D gait analysis.

Anthropometric Comparisons

The anthropometric and body segment parameter data for SI are shown in Tables 5.1 and 5.2. Though SI's height (1.70 m) and mass (69.5 kg) are normal for a young man, Table 5.1 shows some of the anthropometric measures that are a manifestation of his underlying pathology. Note that the circumferences of his right and left calves are 0.292 m and 0.319 m, respectively. The asymmetry of the two sides is quite clear, and the muscle atrophy in the calf muscles can be seen by comparing SI's circumferences with those of the normal male's in Table 3.2. Despite a total body mass of 64.9 kg, almost 5 kg less than that of SI, the normal male shows calf circumferences of 0.365 m. These discrepancies in the APM measurements also translate into differences for the BSP data (cf. Tables 5.2 and 3.3). In the clinical example, the masses of SI's right and left calves are 2.67 kg and 2.85 kg, respectively. Compare these figures with a normal man's masses of 3.28 kg. Although there are fairly small differences for the moment of inertia data of the calves about the flexion/extension and abduction/adduction axes, there are substantial differences about the internal/external axes (cf. 0.0025 kg $\cdot$ m^2 for SI's right calf to 0.0037 kg $\cdot$ m^2 for a normal man's calf). Again, these differences are explained by the atrophy in SI's calf musculature.

Table 5.1 **The Anthropometric Measurements for the Clinical Subject SI (Adult With Cerebral Palsy)**

Number	Anthropometric measurement	Value	Units
1	Total body mass	69.50	kg
2	ASIS breadth	0.250	m
3	R. Thigh length	0.393	m
4	L. Thigh length	0.415	m
5	R. Midthigh circumference	0.524	m
6	L. Midthigh circumference	0.513	m
7	R. Calf length	0.405	m
8	L. Calf length	0.397	m
9	R. Calf circumference	0.292	m
10	L. Calf circumference	0.319	m
11	R. Knee diameter	0.114	m
12	L. Knee diameter	0.110	m
13	R. Foot length	0.238	m
14	L. Foot length	0.243	m
15	R. Malleolus height	0.065	m
16	L. Malleolus height	0.067	m
17	R. Malleolus diameter	0.072	m
18	L. Malleolus diameter	0.066	m
19	R. Foot breadth	0.084	m
20	L. Foot breadth	0.083	m

Note. The format of this table is the same as that of Table 3.2.

Table 5.2 **Body Segment Parameters Generated From the Anthropometric Data of Table 5.1 Using Gaitmath**

Segment number Segment name	1 R. Thigh	2 L. Thigh	3 R. Calf	4 L. Calf	5 R. Foot	6 L. Foot
Mass (kg)	7.53	7.54	2.67	2.85	0.80	0.79
CG position (ratio: proximal/length)	0.39	0.39	0.42	0.42	0.44	0.44
Moments of inertia (kg•m•m)						
Flx/Ext axis	0.1044	0.1133	0.0462	0.0450	0.0032	0.0033
Abd/Add axis	0.1003	0.1089	0.0472	0.0459	0.0036	0.0037
Int/Ext axis	0.0319	0.0307	0.0025	0.0030	0.0010	0.0010

Note. The format of this table is the same as that of Table 3.3.

Kinematic Comparisons

Next, we can examine some simple kinematic measurements. Table 5.3 shows the cycle time, cadence, stride length, and average speed for our clinical example and also for a normal man. When interpreting these data, bear in mind that the man was walking quite slowly. This highlights the fact that the normal range can be fairly large. In the case of our clinical example, SI's cadence of 122.4 steps/min is slightly greater than normal (in fact, it is more like a military pace), but his stride length of 1.22 m is quite short. This combi-

Table 5.3 **Temporal and Distance Parameters for the Clinical Subject SI and a Normal Male**

Parameter	Clinical subject SI	Normal man
Cycle time (s)	0.98	1.30
Cadence (steps/min)	122.40	92.30
Stride length (m)	1.22	1.28
Average speed (m/s)	1.24	0.99

Note. These data were generated in *GaitLab*, using Gaitplot, with the Plot, Options, and List Parameters selected.

nation of cadence and stride length yields a jerky type of steppage gait. Though this cannot be easily interpreted from the data in Table 5.3, it is readily apparent when viewed within *GaitLab*.

Figure 5.2a clearly illustrates SI's balance problems, his wide base of support, and the lurching style that characterize his gait. This figure shows the positions of the left and right heels in the mediolateral (Y) direction as a function of the gait cycle. At right heel strike, the left and right heels have almost the same Y position—0.22 m. Then as the left foot leaves the ground, it swings out laterally to the left, almost to 0.4 m. At left heel strike, the Y position of the heel is 0.37 m and the right heel is at 0.22 m, providing a 0.15 m base of support. Compare this with the base of a normal male in Figure 5.2b, where the heels are never more than 0.10 m apart. At left heel strike, the position of SI's right heel moves about 0.07 m medially. Then at right toe-off, the heel swings out laterally to the right in preparation for the next right heel strike. By viewing SI's gait within *GaitLab*, you will get a clear picture of this in-and-out movement of the feet shown in Figure 5.2a. The normal male illustrated in Figure 5.2b seems to be moving very slightly from left to right as he progresses forward. (Be aware that the curves in Figure 5.2 are based on raw, unfiltered data.)

EMG Comparisons

Figures 5.3 to 5.5 provide a comparison of the EMG for our clinical subject, SI, and a normal man. Before discussing these figures, however, it is important to reiterate that the normal data were not captured from a subject but are based on Winter (1987), whereas the EMG data were gathered directly from SI. You should also be aware that it is very difficult to compare magnitudes between the two sets of graphs (i.e., the EMG values in microvolts, or μV), but a comparison of the phasic activity (i.e., the timing of muscle actions) is entirely valid.

From the sagittal plane data in Figure 5.3a, it is evident that the clinical subject's right knee goes into slight recurvatum during the midstance phase. (This will be discussed later.) Although the EMG patterns for the gluteus maximus are very similar, SI's erector spinae muscle on the right side lacks the biphasic pattern that we see for the normal man in Figure 5.3b. The normal increase in activity during the second double support phase (from left heel strike to right toe-off) is missing.

One of the purposes of the erector spinae (as its name suggests) is to keep the trunk upright; it also helps to stabilize the pelvis as weight is transferred from one leg to the other. Though Figure 5.3b does not show left toe-off (LTO), it is reasonable to assume that a normal man would have good

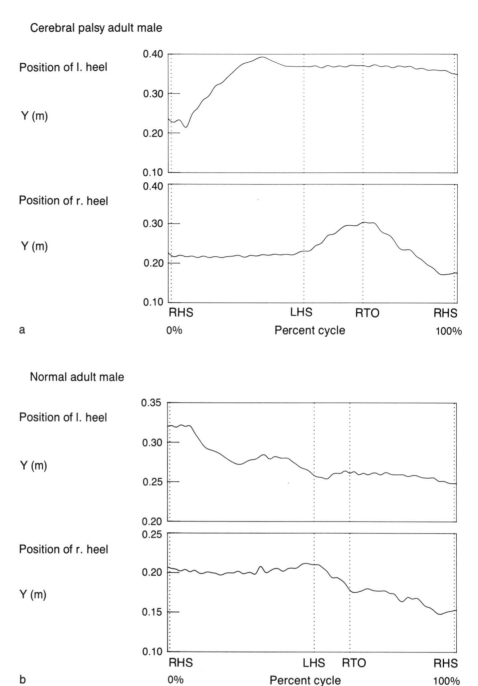

Figure 5.2 The mediolateral position (in meters, m) of the left and right heels as a function of the gait cycle: (a) clinical subject SI; (b) normal man. RHS, right heel strike; LHS, left heel strike; RTO, right toe-off.

left/right symmetry. As a general rule, then, the activity in both the left and right erector spinae rises and falls during the double support phase. The EMGs from these two muscles are in perfect phase with one another (cf. Figure 4.12 and accompanying discussion in chapter 4). Because we did not gather EMG for SI's left side, we cannot say what the activity of that side's erector spinae might be.

Figure 5.4, a and b, compares the EMG activity for the gluteus medius, hamstrings, and rectus femoris muscles. The most obvious discrepancy is in the activity of the gluteus medius during early stance, the loading response phase, when the contralateral foot (here, the left) is about to leave the ground. The sudden decrease in EMG activity in gluteus medius, a hip abductor,

Figure 5.3 Sagittal plane stick figure, muscle activity (in microvolts, μV) of the erector spinae and gluteus maximus plotted as a function of the gait cycle: (a) clinical subject SI; (b) normal man. RHS, right heel strike; LHS, left heel strike; RTO, right toe-off. *Note.* Normal data based on data from Winter (1987).

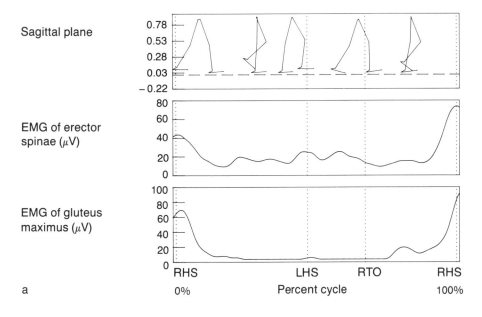

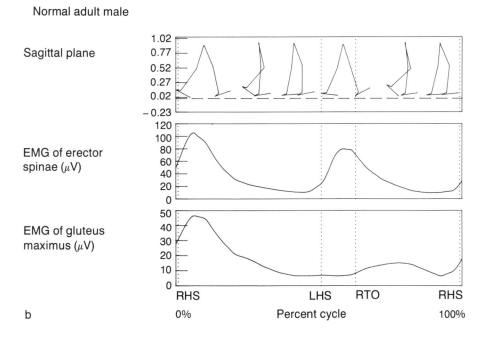

is almost certainly compensated for by the side-to-side lurching described earlier.

The other discrepancy in Figure 5.4 is one of timing in which SI's hamstrings reach their maximum activity level in midswing rather than just before and after heel strike as in a normal male. SI's hamstrings may be trying to control or decelerate the forward jerky motion at the hip (flexion) and knee (extension). The primary role of the hamstrings in normal gait is to stabilize the hip and knee at heel strike.

Figure 5.5, a and b, compares the EMG activity for the adductor longus, tibialis anterior, and triceps surae muscles. Though SI's adductor longus seems a little dysphasic (its highest activity level is during late stance, rather

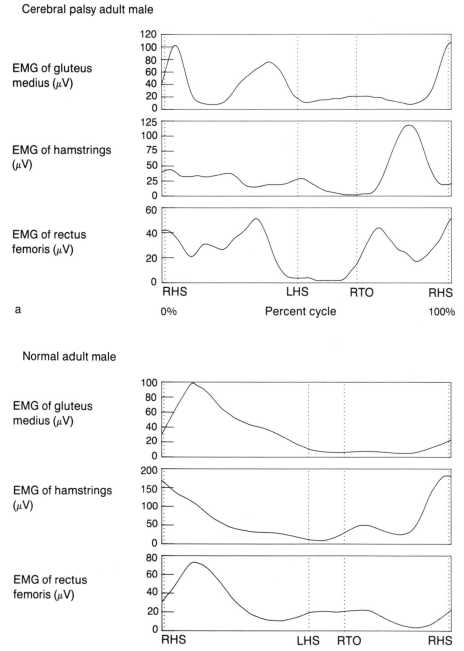

Figure 5.4 Muscle activity (in microvolts, μV) of the gluteus medius, hamstrings, and rectus femoris plotted as a function of the gait cycle: (a) clinical subject SI; (b) normal man. RHS, right heel strike; LHS, left heel strike; RTO, right toe-off. *Note.* Normal data based on data from Winter (1987).

than at toe-off), it is the muscles of the lower leg that show major discrepancies in activity when compared to normal levels. Although you have to be careful when comparing activity levels from two muscles in the same person, it is noteworthy that the EMG for a normal man's tibialis anterior reaches almost 400 μV just after heel strike. In contrast, SI's tibialis anterior is active throughout most of the gait cycle, but its activity level is approximately the same as that of his other muscles. Preoperatively, the neurophysiologist described this muscle as having "scarce, denervated-type muscle potentials." Despite the correction of the bony deformities (seen in Figure 5.1), the underlying neural control of this muscle has probably not been changed. This point

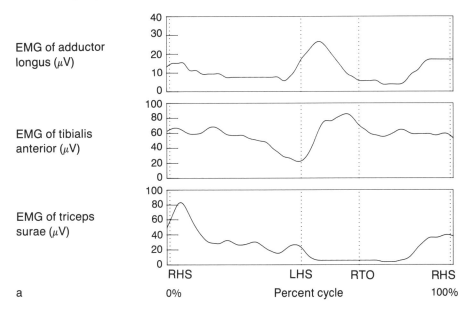

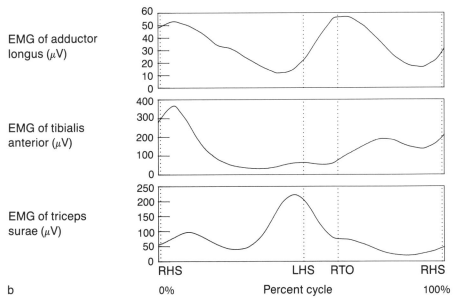

Figure 5.5 Muscle activity (in microvolts, μV) of adductor longus, tibialis anterior, and triceps surae plotted as a function of the gait cycle: (a) clinical subject SI; (b) normal man. RHS, right heel strike; LHS, left heel strike; RTO, right toe-off. *Note.* Normal data based on data from Winter (1987).

was highlighted with regard to the movement chain described in chapter 1 (cf. Figure 1.2 and accompanying discussion).

Perhaps the most interesting feature of Figure 5.5 is the complete lack of triceps surae signal during the pushoff phase (i.e., before left heel strike and right toe-off). Although lacking this pushoff function, which is characteristic of normal gait, SI's triceps surae are active before and after heel strike. Nevertheless, you can recognize SI's gait without seeing him because you can hear the characteristic slapping sound of his feet. This is almost certainly a result of the weakness in the tibialis anterior and its inability to control the plantar flexion that occurs just after heel strike.

Force Plate Comparisons

Figure 5.6, a and b, shows the vertical ground reaction forces experienced by the right (Plate 1) and left (Plate 2) feet of our clinical subject, SI, and a normal man. SI's patterns are not much different from normal. Perhaps the only real clue that these force patterns are the result of some underlying pathology are the exaggerated dips in the curves during midstance. However, as indicated by Vaughan, du Toit, and Roffey (1987b), this pattern could also be produced by a normal person walking at a fast pace. We must emphasize again that the normal subject in Figure 5.6b was walking quite slowly; thus his vertical ground reaction forces are quite flat during midstance.

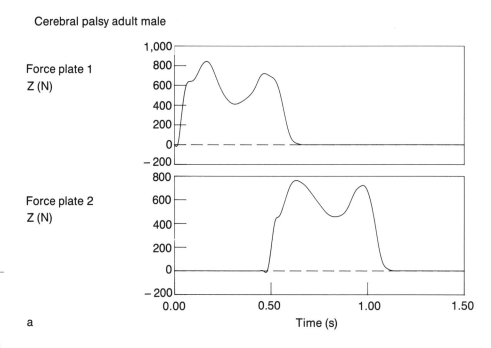

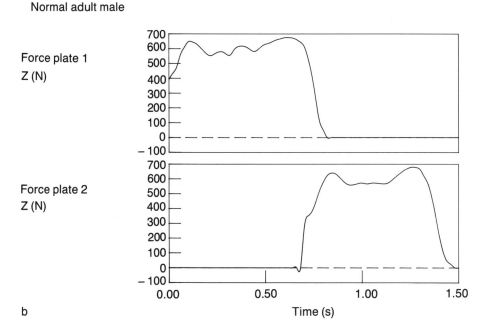

Figure 5.6 Vertical ground reaction forces (in newtons, N) on the right (Plate 1) and left (Plate 2) feet plotted as a function of time: (a) clinical subject SI; (b) normal man. Z is a force in the Z direction.

Ankle Joint Comparisons. In chapter 1 we stated that gait analysts should try to work as far up the movement chain as possible. We also stressed that the key to understanding a person's gait is to integrate the actions at different levels: muscular, joint, and kinematic. This is what we have done in Figures 5.7 to 5.9. In Figure 5.7, a and b, we concentrate on the ankle joint. Notice that the plantar flexion and dorsiflexion angles are very similar; the only difference (and this is slight) is that SI strikes the ground with his foot in about 10° of plantar flexion, hence the slapping action described earlier. The moments at the ankle joint are remarkably similar, with a maximum plantar flexor moment of about 75 N • m during the second double support in both

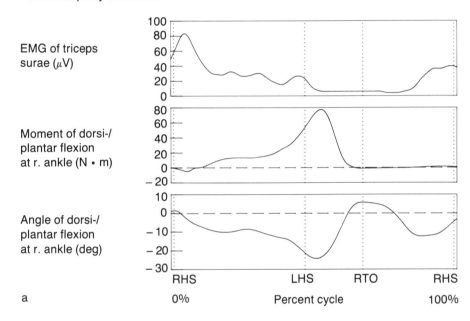

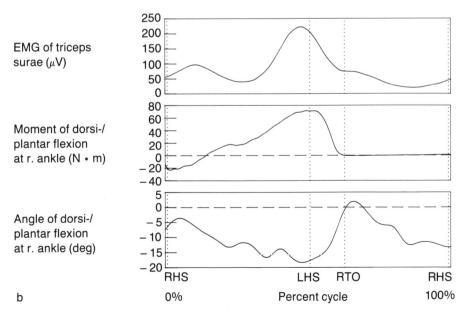

Figure 5.7 Integration of different levels in the movement chain acting about the ankle joint. EMG (in microvolts, μV) of triceps surae, net ankle joint moment (in newton-meters, N•m), and ankle angle (in degrees), plotted as a function of the gait cycle: (a) clinical subject SI; (b) normal man. RHS, right heel strike; LHS, left heel strike; RTO, right toe-off.

cases. What may at first seem puzzling is this: The plantar flexor moment for SI is at its maximum at the same time that the plantar flexor muscle group, the triceps surae, is almost quiescent. How can this apparent anomaly be resolved?

Because of SI's lurching style of gait, he is able to "throw" his body forward, thus ensuring that the ground reaction force passes in front of his ankle joint. As you recall from Equation 3.29, the moment at the ankle joint is largely determined by the ground reaction force. If this force is anterior to (i.e., in front of) the ankle joint, it will tend to dorsiflex the foot. To overcome this action, a plantar flexor moment is generated about the ankle joint. Despite the weakness in the triceps surae, the effect of the arthrodesis is to prevent a collapse of the subtalar joint, thus assisting with the production of a plantar flexor moment.

Knee Joint Comparisons. In Figure 5.8, a and b, we concentrate on the knee joint. At first glance, the curves for SI and a normal male appear to be quite similar. However, on closer examination you will see that in midstance (just before left heel strike), the knee goes into recurvatum (about 10° of extension), the knee moment changes from extension to flexion, and the rectus femoris, a knee extensor, shows its maximum activity. It appears that the knee is compensating in some way for the deficiencies that SI experiences more distally around his ankle joint. With the knee snapping back into hyperextension, the rectus femoris is actively assisting this movement, thus endangering the posterior capsule of the knee. When the knee moment is in flexion, the ground reaction force is acting in an anterior direction. This force is trying to push the knee into further hyperextension, and to overcome this tendency, a net flexor moment must be exerted across the joint.

In Figure 5.4a, when the rectus femoris reaches its maximum value of about 50 μV, the hamstrings are at a relatively low level—about 25 μV. As pointed out previously, you must be very careful when trying to compare these two numbers. However, a net flexor moment indicates that the flexor muscles are dominant over the extensor muscles. It is possible that the surface electrodes placed over the rectus femoris picked up a signal from the whole quadriceps group, whereas the electrodes for the hamstrings were placed laterally over the long head of the biceps femoris, excluding the contributions from semimembranosus and semitendinosus.

It is interesting that Sutherland (1984), in his book *Gait Disorders in Childhood and Adolescence*, presented preoperative data on an 11-year-old girl with Charcot-Marie-Tooth disease. She too had knee recurvatum, clasping of the toes (cf. Figure 5.1), and ground reaction forces that were quite similar to normal except that the center of pressure was concentrated in the forefoot region. Unfortunately, Sutherland did not have any EMG or joint moment data on this patient.

Hip Joint Comparisons. Finally, in Figure 5.9, a and b, we concentrate on the hip joint. The general shape of the curves for gluteus maximus EMG, resultant hip joint moment, and hip angle are very similar for SI and a normal male. The only major difference is the magnitude of SI's flexion moment during pushoff: At 130 N · m, it is almost double a normal male's value. As with the knee joint, it is likely that this high value is in response to the problems SI has distally and to his lurching style of gait.

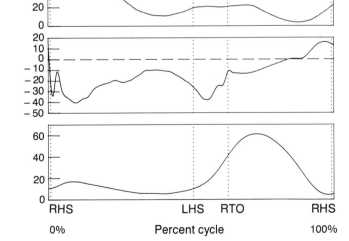

Figure 5.8 Integration of different levels in the movement chain acting about the knee joint. EMG (in microvolts, μV) of rectus femoris, net knee joint moment (in newton-meters, N•m), and knee angle (in degrees), plotted as a function of the gait cycle: (a) clinical subject SI; (b) normal man. RHS, right heel strike; LHS, left heel strike; RTO, right toe-off.

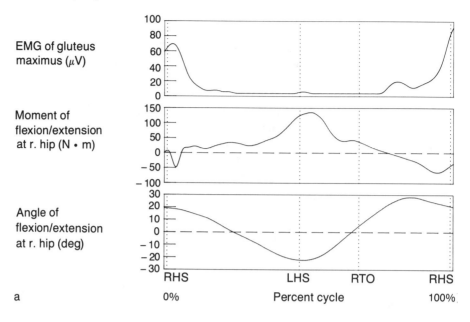

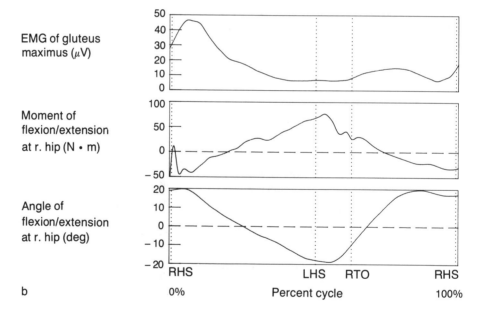

Figure 5.9 Integration of different levels in the movement chain acting about the hip joint. EMG (in microvolts, μV) of gluteus maximus, net hip joint moment (in newton-meters, N•m), and hip angle (in degrees), plotted as a function of the gait cycle: (a) clinical subject SI; (b) normal man. RHS, right heel strike; LHS, left heel strike; RTO, right toe-off.

Summary

When you consider that Figures 5.2 to 5.9 are just a few of the many hundreds of curves that can be generated in *GaitLab*, you realize the potential gait analysis has to aid your understanding of movement dysfunction. We encourage you to explore SI's data in more detail to elucidate both his limitations and his strengths.

APPENDIX A
Dynamic Animation Sequences

> There are two sets of animation sequences—one runs from pages 77 to 131, and the other runs in reverse order from pages 132 to 78. Each includes one complete gait cycle, and they are identical except that the first sequence emphasizes the sagittal view, and the second emphasizes the posterior view. To animate the sequences, isolate the appendixes and place your thumb next to Frame 1 of the sequence you wish to view. Flex the book gently, then fan the pages so that the subject begins to walk. With a little practice you should be able to generate a smooth animation in either direction.

The science of gait analysis has emerged due to the inability of the human eye to measure objectively the many interrelated components of the locomotion system (Brand & Crowninshield, 1981). The word *analysis* comes from the Greek *analyein* and means "to break up" (*Webster's Ninth New Collegiate Dictionary*, 1983). This is precisely what the discipline involves: using measurement techniques to separate kinematic, kinetic, and other parameters describing certain aspects of locomotion. This procedure has its merits for specific applications such as the evaluation of a new surgical technique: By using the tools of gait analysis one can objectively compare specific facets of a patient's walking pattern before and after surgery. In this way, it is possible to avoid making comparisons based only on subjective ideas and recollections of how the patient used to walk.

Unfortunately, by its very nature, gait analysis tends to reduce a dynamic, complex locomotion sequence to a static form, usually graphs and tables. This prevents the human eye from using its marvelous capabilities of global pattern recognition. This human trait, which allows a farmer to spot an injured animal in a herd and enables a boardsailor to select the "perfect wave," has defied attempts at duplication by computer technology. In the medical setting, the computer-generated printout accompanying

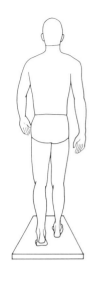
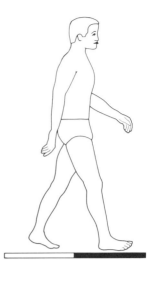
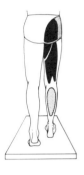
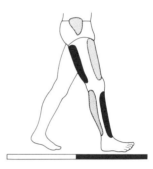

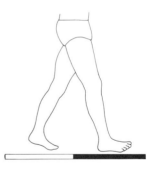

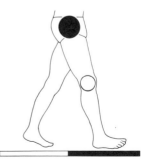

Frame = 1
Time = 0.00 s
Right heel strike

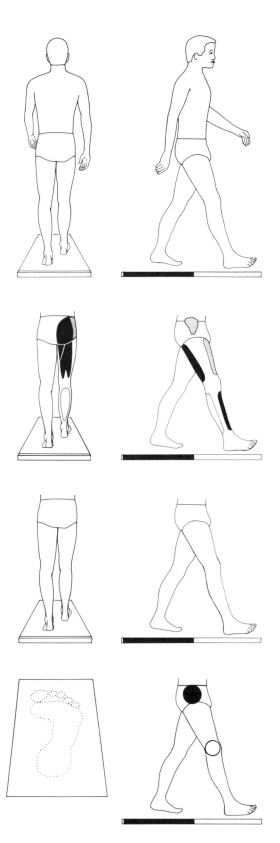

Frame = 28
Time = 1.08 s

an ECG trace is regarded only as a guide—the final decision concerning a heart abnormality is left to the cardiologist, who scans the patterns embedded in the ECG waveform.

Leonardo da Vinci once urged artists to make a "graceful counterbalancing and balancing in such a way that the figure shall not appear as a piece of wood" (Keele, 1983, p. 174). In 1872 Leland Stanford, governor of California, hired Eadweard Muybridge to investigate the question of "unsupported transit": whether a trotting horse ever has all four feet off the ground. For his initial studies, Muybridge took a sequence of photographs, up to 25 pictures/s, that showed that the horse is completely airborne for certain time periods, which allegedly helped Stanford to win a $25,000 wager (Solomon, 1989)! From this early start, Muybridge continued his photographic analyses of the movements of humans and animals at the University of Pennsylvania. His efforts culminated in two classic volumes, *Animals in Motion* (1899) and *The Human Figure in Motion* (1901).

Early Attempts at Integrating Animation and Gait Analysis

Although sequential images contain significant material for all biomechanics researchers, the graphic nature of a printed book cannot convey the feeling of motion. This was recognized by the great French physiologist Etienne Jules Marey who used some of the Muybridge sequences as strips for a zoetrope. Also known as "the wheel of life," the zoetrope had been invented in 1834 by William Horner of Bristol, England: A revolving drum with slits in its sides exploited the principle of the persistence of vision to simulate movement (Solomon, 1989). More recently, Cavanagh (1988), in accepting the Muybridge Medal from the International Society of Biomechanics, animated some of the Muybridge sequences to illustrate his own work on locomotion. In 1990, the Addison Gallery in Andover, MA, took all of Muybridge's sequences and recorded them onto a videodisk. This disk, together with its educational software, has recently been made commercially available by the Voyager Company of Santa Monica, CA, and should provide students of human movement with an outstanding learning resource.

Contributions to Our Animation Studies

Our own efforts to use animation in understanding the mechanics of human gait have been influenced not only by Muybridge, but by some recent workers, too. An important contribution was made by Inman et al. (1981), who introduced a novel method for illustrating the actions of leg muscles during the gait cycle (refer to the discussion in chapter 4). Cavanagh, Hennig, Bunch, and Macmillan (1983) measured the pressure distribution beneath the plantar surface of the foot during walking and animated wire-frame diagrams of the pressure profile, using a graphics display computer and a movie camera; as the pressure "flowed" from the rear of the foot to the front of the foot, the perspective also changed, offering the observer some additional insights. As one of the appendixes to his PhD thesis, van den Bogert (1989) included a software diskette that enabled a user to run animated sequences of a walking horse on a standard personal computer with CGA graphics. These sequences, which were generated from a computer simulation package, could be run at various speeds.

Reversal of Gait Analysis

Beginning in this appendix, and continuing into the next, the process of gait analysis will be reversed and a single, static figure will be brought to life. This will be done by having you, the reader, fan the pages of the appendixes to animate our still-frame images. These animation sequences will help you gain a new appreciation of the human trademark, bipedality. Note that though it is possible to make the figure walk backward, the videotape that formed the basis for these animation sequences was filmed at 25 frames/s for a total of 28 frames/cycle and a normal, forward walking pace of 1.7 m/s on a treadmill.

There were two primary sources for the data superimposed on the animation sequences. These were Winter (1987) for the joint moments, electromyography, and ground reaction forces, and the Center for Locomotion Studies (CELOS) at Pennsylvania State University for the plantar pressure profiles.

Note that whereas the edges of the book have animation figures for both the sagittal and frontal planes, the computer animation in GaitLab shows only the sagittal view.

Sagittal Plane Motion

The four figures showing a right, sagittal plane view of the person indicate, from top to bottom,

- total body motion;
- muscular activation of gluteus medius, quadriceps, hamstrings, tibialis anterior, and triceps surae;
- ground reaction forces; and
- resultant joint moments.

Note that the joint moments are based on the inverse dynamics approach (Winter, 1987). The solid circles represent extension and dorsiflexion moments, whereas hollow circles represent flexion and plantar flexion moments. The radius of each circle is proportional to the magnitude of the corresponding moment, with the three joints being plotted to the same scale (cf. Figures 5.7b, 5.8b, and 5.9b). A similar, icon-based approach to illustrating moments has recently been suggested by Loeb and Levine (1990). Greaves (1990) has also recently demonstrated software that overlays the 3-D ground reaction force and joint moment as a vector on an animated stick figure. We would have liked to include more muscles in our animation sequences (e.g., gluteus maximus, a hip extensor). However, we felt that there was a limit to the amount of information that could be displayed without crowding the images unnecessarily. Therefore, we must

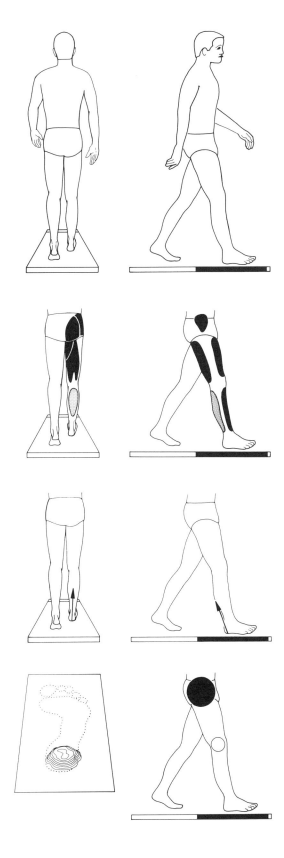

Frame = 2
Time = 0.04 s

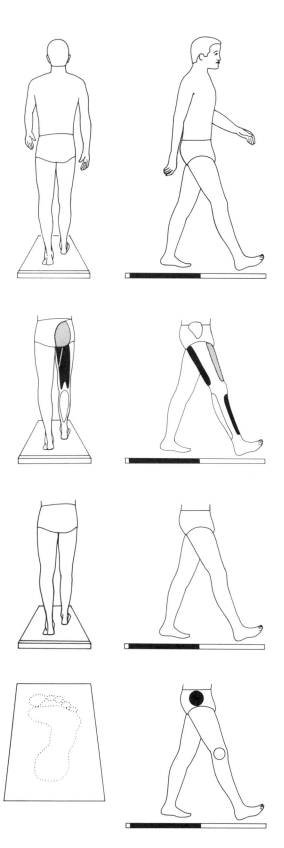

Frame = 27
Time = 1.04 s

emphasize that some muscles, vitally important in human gait, have not been included here purely for reasons of space.

It is said that a picture is worth a thousand words, and this is certainly true for this animation sequence. In particular, pay attention to the following:

- The relationship between ground reaction force and joint moments—the figures illustrate the error that can be made by assuming a quasi-static situation. If the inertial contributions to joint moment were neglected, there would be no moments during the swing phase, and hip moments during stance would exhibit a completely different behavior. In Frame 13, for instance, if the hip moment had been estimated using the ground reaction force alone, it would have had an opposite value to its true extensor moment.
- The relationship between joint moment and muscle activation—in particular note the stabilizing role of the muscles in counteracting the moment applied to the joints.
- The orientation of the ground reaction force—at first, the ground reaction force acts posteriorly, but later, during pushoff, it acts anteriorly.
- The ratio between stance and swing times—typically 60%:40% of the gait cycle.

Frontal Plane Motion

The four figures showing a posterior, frontal plane view of the person illustrate, from top to bottom

- total body motion;
- muscular activation of gluteus maximus, gluteus medius, adductor magnus, hamstrings, and gastrocnemius;
- ground reaction forces; and
- pressure distribution on the plantar surface of the foot.

We chose not to include frontal plane joint moments because we felt that they would clutter the figures. A few key muscles, such as tibialis anterior, were not included for the same reason. The following are points that should be noted:

- The slight, side-to-side movement of the torso as first one foot then the other is lifted off the ground. In the case of a person with weak hip abductors, this movement is far more pronounced.
- The phasic behavior and stabilization role of the muscles (e.g., gluteus medius activation at contralateral toe-off)
- The relationship between ground reaction force and pressure distribution under the foot (the contour intervals are 25 kPa, or kilopascals)
- The transfer of weight from the heel across the foot's lateral border on to the metatarsal heads, and the important role of the toes during pushoff

It may take a while to become proficient at flipping backward and forward through these animation sequences, but when mastered, it should take about 1 s to visualize each complete gait cycle. Do not assume that by flipping through faster you can get an impression of speed walking—a person's entire stride pattern, muscle activation, ground reaction forces, and plantar pressure profile change with a change in speed. However, at least in this particular sequence, a normal male's gait can be visualized using pattern recognition, unobscured by stick figures or tables of data that would be understood only by those actively engaged in locomotion research.

Summary

The animation of human gait is certainly not new, having been introduced by the pioneering studies of Muybridge a century ago. What *is* new is the integration of the animated movement with the underlying causes of human gait: muscle activity, joint moments, and ground reaction forces. We believe that the sequences on these pages will bring the walking subject alive, providing you with new insights; also, we encourage you to spend time with *GaitLab*'s Animate software.

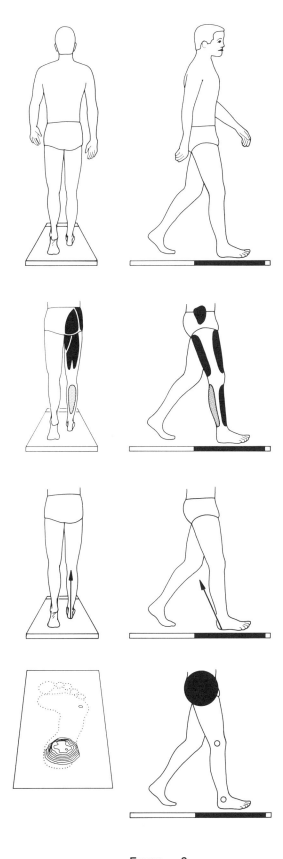

Frame = 3
Time = 0.08 s
Right foot-flat

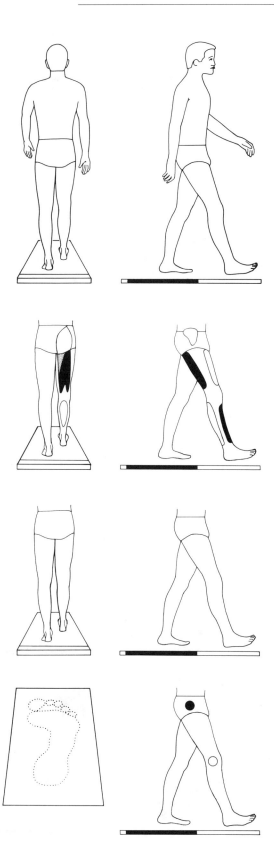

Frame = 26
Time = 1.00 s
Right deceleration

APPENDIX B
Detailed Mathematics Used in Gaitmath

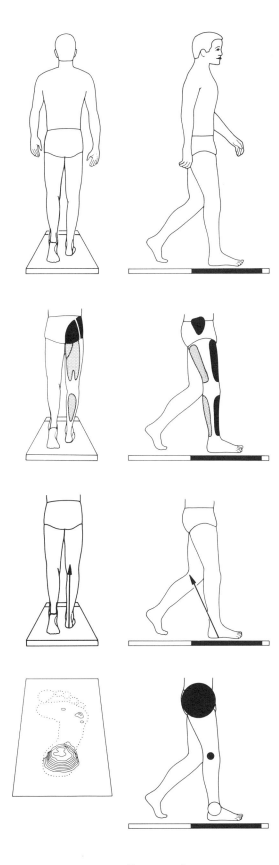

Frame = 4
Time = 0.12 s

This appendix contains the detailed mathematics that are used to process the anthropometric, kinematic, and force plate data files. These details have been incorporated in the *GaitLab* program Gaitmath. Because we do not provide a listing of the source code for Gaitmath, and because the material presented in chapter 3 tends to gloss over many details, we have provided all the necessary details for researchers of human gait in this appendix.

Like chapter 3, this appendix covers five different topics: body segment parameters; linear kinematics; centers of gravity; angular kinematics; and dynamics of joints.

Body Segment Parameters

We have chosen to use a method for predicting body segment parameters that is based on simple geometric modeling combined with the anthropometric data of Chandler et al. (1975). The thighs and calves are modeled by right rectangular cylinders, whereas the feet are modeled by right rectangular pyramids. The key point to bear in mind is that our modeling process makes use of *dimensional* consistency. By this we mean that only parameters that have the composite units of kilograms are used to predict segment mass, and that only parameters with the composite units of kg • m^2 are used to predict segmental moments of inertia. We believe, for example, that it makes little sense to use only total body mass to predict segmental moments of inertia. (This was the method used by Chandler et al., 1975.) We will show later in this section how much better our method is in predicting segment moments of inertia.

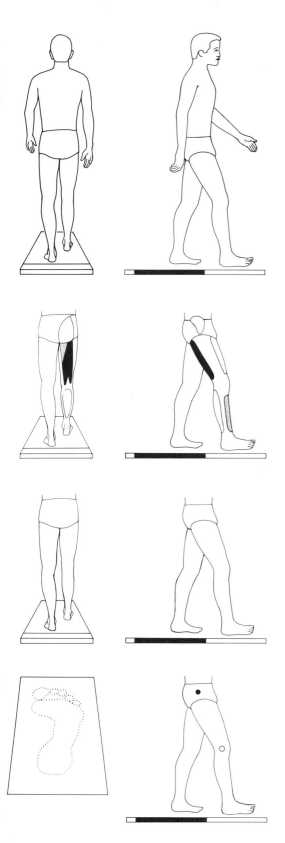

Equations 3.1 to 3.3 describe the format and rationale for generating regression equations to predict segment mass based on anthropometric data. The relevant parameters, A_1 through A_{20}, are presented in Table B.1. (A description of how to make these measurements is provided in Table 3.1.) The regression equations that we derived (Equations 3.4 to 3.6 in chapter 3) are based on the six cadavers in Chandler et al. (1975) and are repeated here, for sake of completeness, in Table B.2. This table also lists the center of gravity ratios, which are based on the mean values of the cadavers.

In Equations 3.7 to 3.10 and Figure 3.3 we argued for regression equations to predict segmental moments of inertia that are based on body mass in kilograms (kg) times a composite parameter having the dimensions of length squared (m^2). Equation 3.11 was presented as one example (in this case, for the moment of inertia of the thigh about the flexion/extension axis) of such a regression equation. The full set of 18 equations (right and left thighs, calves, and feet, about their flexion/extension, abduction/adduction, internal/external rotation axes) is presented in Table B.3. In trying to understand the relevant axes, refer to Figures 3.3 and 3.10 and the following key:

$$\begin{aligned} \text{FlxExt} &= z \text{ axis} \\ \text{AbdAdd} &= y \text{ axis} \\ \text{IntExt} &= x \text{ axis} \end{aligned}$$

Table B.1 Anthropometric Data for Calculating Body Segment Parameters and for Predicting Joint Centers and Segment Endpoints

Parameter number	Name
A_1	Total body mass
A_2	Anterior superior iliac spine (ASIS) breadth
A_3	Right thigh length
A_4	Left thigh length
A_5	Right midthigh circumference
A_6	Left midthigh circumference
A_7	Right calf length
A_8	Left calf length
A_9	Right calf circumference
A_{10}	Left calf circumference
A_{11}	Right knee diameter
A_{12}	Left knee diameter
A_{13}	Right foot length
A_{14}	Left foot length
A_{15}	Right malleolus height
A_{16}	Left malleolus height
A_{17}	Right malleolus width
A_{18}	Left malleolus width
A_{19}	Right foot breadth
A_{20}	Left foot breadth

Frame = 25
Time = 0.96 s

Table B.2 Equations to Predict the Masses and Centers of Gravity for the Thigh, Calf, and Foot

Mass.R.Thigh := (0.1032) • A1 + (12.76) • A3 • A5 • A5 − 1.023;
Mass.L.Thigh := (0.1032) • A1 + (12.76) • A4 • A6 • A6 − 1.023;
Mass.R.Calf := (0.0226) • A1 + (31.33) • A7 • A9 • A9 + 0.016;
Mass.L.Calf := (0.0226) • A1 + (31.33) • A8 • A10 • A10 + 0.016;
Mass.R.Foot := (0.0083) • A1 + (254.5) • A13 • A15 • A17 − 0.065;
Mass.L.Foot := (0.0083) • A1 + (254.5) • A14 • A16 • A18 − 0.065;

CG_Ratio.R.Thigh := 0.39;
CG_Ratio.L.Thigh := 0.39;
CG_Ratio.R.Calf := 0.42;
CG_Ratio.L.Calf := 0.42;
CG_Ratio.R.Foot := 0.44;
CG_Ratio.L.Foot := 0.44;

Note. A1 through A18 are the anthropometric parameters defined in Table B.1. The format of these equations is exactly the same as the Turbo Pascal source code in Gaitmath.

Table B.3 Equations to Predict Moments of Inertia (I) for the Thigh, Calf, and Foot

I_FlxExt.R.Thigh := 0.00762 • A1 • (A3 • A3 + 0.076 • A5 • A5) + 0.01153;
I_FlxExt.L.Thigh := 0.00762 • A1 • (A4 • A4 + 0.076 • A6 • A6) + 0.01153;
I_AbdAdd.R.Thigh := 0.00726 • A1 • (A3 • A3 + 0.076 • A5 • A5) + 0.01186;
I_AbdAdd.L.Thigh := 0.00726 • A1 • (A4 • A4 + 0.076 • A6 • A6) + 0.01186;
I_IntExt.R.Thigh := 0.00151 • A1 • A5 • A5 + 0.00305;
I_IntExt.L.Thigh := 0.00151 • A1 • A6 • A6 + 0.00305;

I_FlxExt.R.Calf := 0.00347 • A1 • (A7 • A7 + 0.076 • A9 • A9) + 0.00511;
I_FlxExt.L.Calf := 0.00347 • A1 • (A8 • A8 + 0.076 • A10 • A10) + 0.00511;
I_AbdAdd.R.Calf := 0.00387 • A1 • (A7 • A7 + 0.076 • A9 • A9) + 0.00138;
I_AbdAdd.L.Calf := 0.00387 • A1 • (A8 • A8 + 0.076 • A10 • A10) + 0.00138;
I_IntExt.R.Calf := 0.00041 • A1 • A9 • A9 + 0.00012;
I_IntExt.L.Calf := 0.00041 • A1 • A10 • A10 + 0.00012;

I_FlxExt.R.Foot := 0.00023 • A1 • (4 • A15 • A15 + 3 • A13 • A13) + 0.00022;
I_FlxExt.L.Foot := 0.00023 • A1 • (4 • A16 • A16 + 3 • A14 • A14) + 0.00022;
I_AbdAdd.R.Foot := 0.00021 • A1 • (4 • A19 • A19 + 3 • A13 • A13) + 0.00067;
I_AbdAdd.L.Foot := 0.00021 • A1 • (4 • A20 • A20 + 3 • A14 • A14) + 0.00067;
I_IntExt.R.Foot := 0.00141 • A1 • (A15 • A15 + A19 • A19) − 0.00008;
I_IntExt.L.Foot := 0.00141 • A1 • (A16 • A16 + A20 • A20) − 0.00008;

Note. A1 through A20 are the anthropometric parameters defined in Table B.1. The format of these equations is exactly the same as the Turbo Pascal source code in Gaitmath.

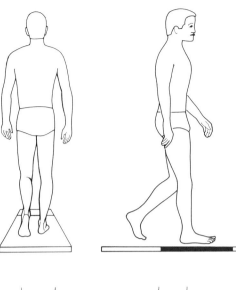
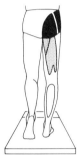
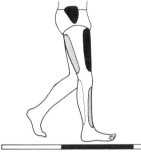

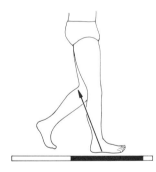

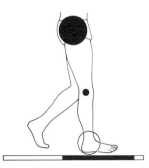

Frame = 5
Time = 0.16 s
Left toe-off

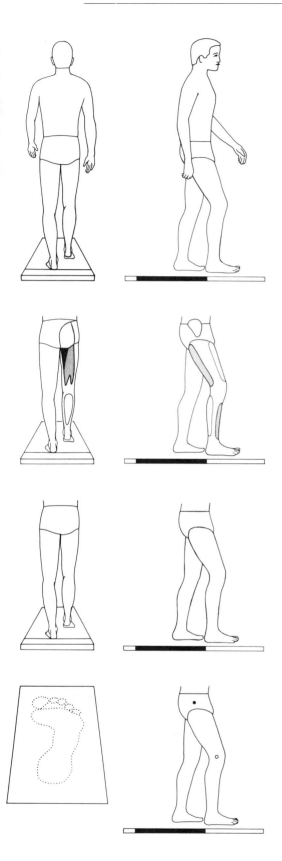

Frame = 24
Time = 0.92 s

Because we have proposed that a gait analyst should take the time to measure 20 anthropometric parameters (Table B.1) and use these data in our regression equations (Tables B.2 and B.3), it is reasonable to ask, Is there any benefit? We believe that there *is* benefit in personalizing the BSPs. Chandler et al. (1975) derived regression equations based only on total body mass for predicting segmental masses and moments of inertia. Their correlation coefficients, which are a measure of how well their equations fitted the data, are presented in Table B.4. For comparison, our correlation coefficients are also included in this table. Because Equations 3.4 to 3.6 (top of Table B.2) used more than one parameter to predict segment mass (total body mass and a composite parameter representing segment volume), it is necessary to calculate R', the correlation coefficient adjusted to allow for shrinkage:

$$R' = [R^2 - \frac{(p-1)}{N-p}(1 - R^2)]^{\frac{1}{2}} \quad \text{(B.1)}$$

where N is the number of cadavers (6), p is the number of predictors (2), and R is the unadjusted multiple correlation coefficient (Kim & Kohout, 1975). You can see that if $p = 1$ or $N \gg p$, then $R' = R$.

Note that for each of the segment masses, the adjusted coefficient was substantially better than the simple correlation coefficients of Chandler et al. (1975). Note, too, that the correlation coefficients for the moments of inertia equations proposed in the current method were in all cases (except one) markedly

Table B.4 Comparison of Methods Used to Predict Body Segment Parameters for 6 Cadavers of Chandler et al. (1975)

Parameter	Segment	Chandler's method Correlation coefficient	Gaitmath method Correlation coefficient	Gaitmath method Adjusted coefficient[a]
Mass	Thigh	0.941	0.998	0.997
	Calf	0.917	0.997	0.996
	Foot	0.784	0.899	0.872
Moment of inertia	Thigh			
	FlxExt	0.865	0.901	
	AbdAdd	0.939	0.913	
	IntExt	0.876	0.932	
	Calf			
	FlxExt	0.850	0.972	
	AbdAdd	0.821	0.962	
	IntExt	0.795	0.896	
	Foot			
	FlxExt	0.696	0.899	
	AbdAdd	0.762	0.871	
	IntExt	0.819	0.825	

[a]The correlation coefficients for the Gaitmath method have to be adjusted for shrinkage, because the equations to predict segment mass are based on more than one composite parameter. Refer to text for more detail.

higher than those of Chandler. In that one case (the moment of inertia of the thigh about the abduction/adduction axis), our coefficient of 0.913 is still quite acceptable. It was not necessary to calculate an adjusted correlation coefficient for our moments of inertia, because only one predictor—a composite parameter having the dimension kilogram • meter • meter (kg • m^2)—was used.

We believe that the evidence contained in Table B.4 provides encouraging support for our suggestion that the equations in Tables B.2 and B.3 are of benefit to the gait analyst, who can use these equations to personalize the BSPs of a subject knowing that they work extremely well with the original data set on which they were based. Also, because the equations are dimensionally consistent, they can be used with data from subjects whose sizes and shapes may be quite different from those of the 6 male cadavers of Chandler et al. (1975). The equations can be used on children or women or tall basketball players without giving unreasonable answers. The same cannot be said for regression equations, such as those proposed by Hinrichs (1985), that are not dimensionally consistent. This important issue has recently been addressed by Yeadon and Morlock (1989).

Linear Kinematics

In this section we show how the 15 marker positions (see Figure 3.4 and Table B.5) may be used to accomplish two primary tasks. The first is to calculate **uvw** reference systems for each segment (see Figures 3.6 to 3.8) to predict the positions of joint centers and segment endpoints (see Equations 3.13 to 3.16). The second task is to use the joint center positions and the external marker positions (Table B.5) to generate segment reference frames (xyz), which are embedded at the centers of gravity of each segment (see Figure 3.10).

Table B.5 Position Numbers and Names of External Marker Positions

Position number	Name
p_1	Right metatarsal head V
p_2	Right heel
p_3	Right lateral malleolus
p_4	Right tibial tubercle
p_5	Right femoral epicondyle
p_6	Right greater trochanter
p_7	Right anterior superior iliac spine
p_8	Left metatarsal head V
p_9	Left heel
p_{10}	Left lateral malleolus
p_{11}	Left tibial tubercle
p_{12}	Left femoral epicondyle
p_{13}	Left greater trochanter
p_{14}	Left anterior superior iliac spine
p_{15}	Sacrum

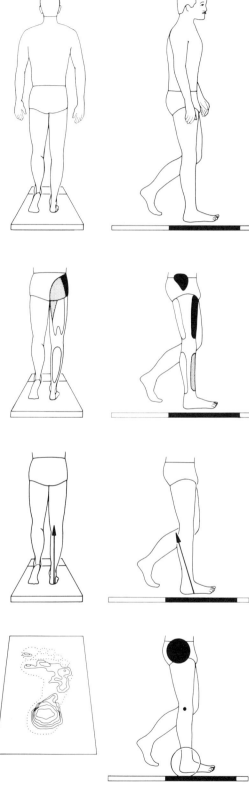

Frame = 6
Time = 0.20 s

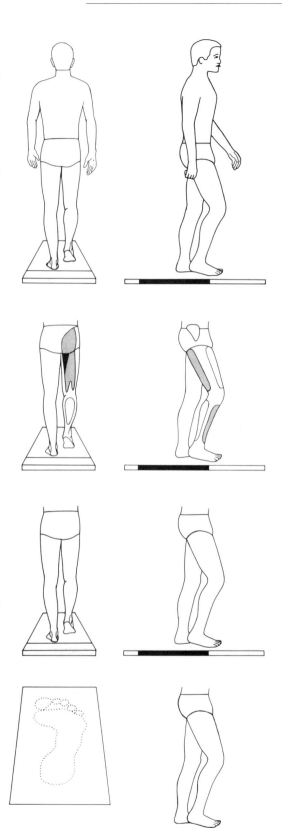

Frame = 23
Time = 0.88 s

From Figure 3.6, we may define the unit vector triad **uvw** for the right foot as follows:

$$\mathbf{u}_{R.Foot} = (\mathbf{p}_1 - \mathbf{p}_2)/|\mathbf{p}_1 - \mathbf{p}_2| \quad (B.2)$$

$$\mathbf{w}_{R.Foot} = \frac{(\mathbf{p}_1 - \mathbf{p}_3) \times (\mathbf{p}_2 - \mathbf{p}_3)}{|(\mathbf{p}_1 - \mathbf{p}_3) \times (\mathbf{p}_2 - \mathbf{p}_3)|} \quad (B.3)$$

$$\mathbf{v}_{R.Foot} = \mathbf{w}_{R.Foot} \times \mathbf{u}_{R.Foot} \quad (B.4)$$

Then, based on stereo X rays (Vaughan, 1983), we have the following equations:

$$\mathbf{p}_{R.Ankle} = \mathbf{p}_3 - 0.008 A_{13}\, \mathbf{u}_{R.Foot} + 0.393 A_{15}\, \mathbf{v}_{R.Foot} \quad (B.5)$$
$$+ 0.706 A_{17}\, \mathbf{w}_{R.Foot}$$

and

$$\mathbf{p}_{R.Toe} = \mathbf{p}_3 + 0.697 A_{13}\, \mathbf{u}_{R.Foot} + 0.780 A_{15}\, \mathbf{v}_{R.Foot} \quad (B.6)$$
$$+ 0.923 A_{19}\, \mathbf{w}_{R.Foot}$$

which are the same as Equations 3.13 and 3.14. Similarly, we may calculate the unit vector triad **uvw** for the left foot as follows:

$$\mathbf{u}_{L.Foot} = (\mathbf{p}_8 - \mathbf{p}_9)/|\mathbf{p}_8 - \mathbf{p}_9| \quad (B.7)$$

$$\mathbf{w}_{L.Foot} = \frac{(\mathbf{p}_8 - \mathbf{p}_{10}) \times (\mathbf{p}_9 - \mathbf{p}_{10})}{|(\mathbf{p}_8 - \mathbf{p}_{10}) \times (\mathbf{p}_9 - \mathbf{p}_{10})|} \quad (B.8)$$

$$\mathbf{v}_{L.Foot} = \mathbf{w}_{L.Foot} \times \mathbf{u}_{L.Foot} \quad (B.9)$$

As before, this unit vector triad may be used to estimate the following:

$$\mathbf{p}_{L.Ankle} = \mathbf{p}_{10} - 0.008 A_{14}\, \mathbf{u}_{L.Foot} + 0.393 A_{16}\, \mathbf{v}_{L.Foot} \quad (B.10)$$
$$- 0.706 A_{18}\, \mathbf{w}_{L.Foot}$$

and

$$\mathbf{p}_{L.Toe} = \mathbf{p}_{10} + 0.697 A_{14}\, \mathbf{u}_{L.Foot} + 0.780 A_{16}\, \mathbf{v}_{L.Foot} \quad (B.11)$$
$$- 0.923 A_{20}\, \mathbf{w}_{L.Foot}$$

which are similar to Equations B.5 and B.6, the main difference being that $\mathbf{w}_{R.Foot}$ points medially, whereas $\mathbf{w}_{L.Foot}$ points laterally.

From Figure 3.7, we may define the unit vector triad **uvw** for the right calf as follows:

$$\mathbf{v}_{R.Calf} = (\mathbf{p}_3 - \mathbf{p}_5)/|\mathbf{p}_3 - \mathbf{p}_5| \quad (B.12)$$

$$\mathbf{w}_{R.Calf} = \frac{(\mathbf{p}_4 - \mathbf{p}_5) \times (\mathbf{p}_3 - \mathbf{p}_5)}{|(\mathbf{p}_4 - \mathbf{p}_5) \times (\mathbf{p}_3 - \mathbf{p}_5)|} \quad (B.13)$$

$$\mathbf{u}_{R.Calf} = \mathbf{v}_{R.Calf} \times \mathbf{w}_{R.Calf} \quad (B.14)$$

We can now calculate the position of the right knee:

$$\mathbf{p}_{R.Knee} = \mathbf{p}_5 + 0.423 A_{11}\, \mathbf{u}_{R.Calf} - 0.198 A_{11}\, \mathbf{v}_{R.Calf} \quad (B.15)$$
$$+ 0.406 A_{11}\, \mathbf{w}_{R.Calf}$$

which is the same as Equation 3.15. Similarly, we may calculate the unit vector triad **uvw** for the left calf as follows:

$$\mathbf{v}_{L.Calf} = (\mathbf{p}_{10} - \mathbf{p}_{12})/|\mathbf{p}_{10} - \mathbf{p}_{12}| \quad (B.16)$$

$$\mathbf{w}_{L.Calf} = \frac{(\mathbf{p}_{11} - \mathbf{p}_{12}) \times (\mathbf{p}_{10} - \mathbf{p}_{12})}{|(\mathbf{p}_{11} - \mathbf{p}_{12}) \times (\mathbf{p}_{11} - \mathbf{p}_{12})|} \quad (B.17)$$

$$\mathbf{u}_{L.Calf} = \mathbf{v}_{L.Calf} \times \mathbf{w}_{L.Calf} \quad (B.18)$$

As before, this vector triad may be used to estimate the position of the left knee:

$$\mathbf{p}_{L.Knee} = \mathbf{p}_{12} + 0.423 A_{12} \mathbf{u}_{L.Calf} - 0.198 A_{12} \mathbf{v}_{L.Calf} \quad (B.19)$$
$$- 0.406 A_{12} \mathbf{w}_{L.Calf}$$

which is similar to Equation B.15, the main difference being that $\mathbf{w}_{R.Calf}$ points medially, whereas $\mathbf{w}_{L.Calf}$ points laterally.

From Figure 3.8, we may define the unit vector triad **uvw** for the pelvis as follows:

$$\mathbf{v}_{Pelvis} = (\mathbf{p}_{14} - \mathbf{p}_7)/|\mathbf{p}_{14} - \mathbf{p}_7| \quad (B.20)$$

$$\mathbf{w}_{Pelvis} = \frac{(\mathbf{p}_7 - \mathbf{p}_{15}) \times (\mathbf{p}_{14} - \mathbf{p}_{15})}{|(\mathbf{p}_7 - \mathbf{p}_{15}) \times (\mathbf{p}_{14} - \mathbf{p}_{15})|} \quad (B.21)$$

$$\mathbf{u}_{Pelvis} = \mathbf{v}_{Pelvis} \times \mathbf{w}_{Pelvis} \quad (B.22)$$

This same vector triad may be used to calculate the positions of both the right and left hips:

$$\mathbf{p}_{R.Hip} = \mathbf{p}_{15} + 0.598 A_2 \mathbf{u}_{Pelvis} - 0.344 A_2 \mathbf{v}_{Pelvis} \quad (B.23)$$
$$- 0.290 A_2 \mathbf{w}_{Pelvis}$$

$$\mathbf{p}_{L.Hip} = \mathbf{p}_{15} + 0.598 A_2 \mathbf{u}_{Pelvis} + 0.344 A_2 \mathbf{v}_{Pelvis} \quad (B.24)$$
$$- 0.290 A_2 \mathbf{w}_{Pelvis}$$

These equations (B.23 and B.24) for predicting the position of the hip joints are very similar to others in the literature (Campbell et al., 1988; Tylkowski et al., 1982).

The next task is to use the joint center positions and external marker positions to generate segment reference frames (xyz), which are embedded at the centers of gravity of each segment (see Figure 3.10). There are a few observations that need to be made first:

IJK are the unit vectors in the XYZ directions;
ijk are the unit vectors in the xyz directions.

$\mathbf{i}_{Pelvis} = \mathbf{w}_{Pelvis}$
$\mathbf{j}_{Pelvis} = \mathbf{u}_{Pelvis}$
$\mathbf{k}_{Pelvis} = \mathbf{v}_{Pelvis}$
Segment 1 is the Right Thigh;
Segment 2 is the Left Thigh;
Segment 3 is the Right Calf;
Segment 4 is the Left Calf;
Segment 5 is the Right Foot;
Segment 6 is the Left Foot.

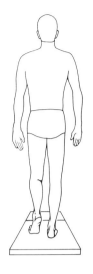
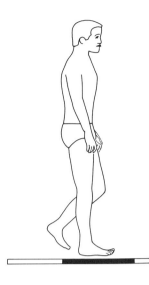

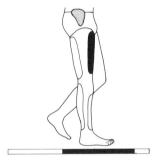

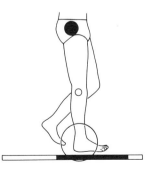

Frame = 7
Time = 0.24 s

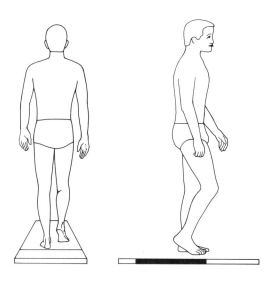

The unit vector triad **ijk** defining the directions of xyz in the segments may be calculated as follows:

Right Thigh.

$$\mathbf{i}_1 = \frac{(\mathbf{p}_{R.Hip} - \mathbf{p}_{R.Knee})}{|\mathbf{p}_{R.Hip} - \mathbf{p}_{R.Knee}|} \quad (B.25)$$

$$\mathbf{j}_1 = \frac{(\mathbf{p}_6 - \mathbf{p}_{R.Hip}) \times (\mathbf{p}_{R.Knee} - \mathbf{p}_{R.Hip})}{|(\mathbf{p}_6 - \mathbf{p}_{R.Hip}) \times (\mathbf{p}_{R.Knee} - \mathbf{p}_{R.Hip})|} \quad (B.26)$$

$$\mathbf{k}_1 = \mathbf{i}_1 \times \mathbf{j}_1 \quad (B.27)$$

Left Thigh.

$$\mathbf{i}_2 = \frac{(\mathbf{p}_{L.Hip} - \mathbf{p}_{L.Knee})}{|\mathbf{p}_{L.Hip} - \mathbf{p}_{L.Knee}|} \quad (B.28)$$

$$\mathbf{j}_2 = \frac{(\mathbf{p}_{L.Knee} - \mathbf{p}_{L.Hip}) \times (\mathbf{p}_{13} - \mathbf{p}_{L.Hip})}{|(\mathbf{p}_{L.Knee} - \mathbf{p}_{L.Hip}) \times (\mathbf{p}_{13} - \mathbf{p}_{L.Hip})|} \quad (B.29)$$

$$\mathbf{k}_2 = \mathbf{i}_2 \times \mathbf{j}_2 \quad (B.30)$$

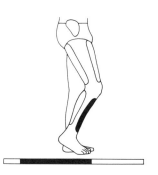

Right Calf.

$$\mathbf{i}_3 = \frac{(\mathbf{p}_{R.Knee} - \mathbf{p}_{R.Ankle})}{|\mathbf{p}_{R.Knee} - \mathbf{p}_{R.Ankle}|} \quad (B.31)$$

$$\mathbf{j}_3 = \frac{(\mathbf{p}_5 - \mathbf{p}_{R.Knee}) \times (\mathbf{p}_{R.Ankle} - \mathbf{p}_{R.Knee})}{|(\mathbf{p}_5 - \mathbf{p}_{R.Knee}) \times (\mathbf{p}_{R.Ankle} - \mathbf{p}_{R.Knee})|} \quad (B.32)$$

$$\mathbf{k}_3 = \mathbf{i}_3 \times \mathbf{j}_3 \quad (B.33)$$

Left Calf.

$$\mathbf{i}_4 = \frac{(\mathbf{p}_{L.Knee} - \mathbf{p}_{L.Ankle})}{|\mathbf{p}_{L.Knee} - \mathbf{p}_{L.Ankle}|} \quad (B.34)$$

$$\mathbf{j}_4 = \frac{(\mathbf{p}_{L.Ankle} - \mathbf{p}_{L.Knee}) \times (\mathbf{p}_{12} - \mathbf{p}_{L.Knee})}{|(\mathbf{p}_{L.Ankle} - \mathbf{p}_{L.Knee}) \times (\mathbf{p}_{12} - \mathbf{p}_{L.Knee})|} \quad (B.35)$$

$$\mathbf{k}_4 = \mathbf{i}_4 \times \mathbf{j}_4 \quad (B.36)$$

Right Foot.

$$\mathbf{i}_5 = \frac{(\mathbf{p}_2 - \mathbf{p}_{R.Toe})}{|\mathbf{p}_2 - \mathbf{p}_{R.Toe}|} \quad (B.37)$$

$$\mathbf{k}_5 = \frac{(\mathbf{p}_{R.Ankle} - \mathbf{p}_2) \times (\mathbf{p}_{R.Toe} - \mathbf{p}_2)}{|(\mathbf{p}_{R.Ankle} - \mathbf{p}_2) \times (\mathbf{p}_{R.Toe} - \mathbf{p}_2)|} \quad (B.38)$$

$$\mathbf{j}_5 = \mathbf{k}_5 \times \mathbf{i}_5 \quad (B.39)$$

Left Foot.

$$\mathbf{i}_6 = \frac{(\mathbf{p}_9 - \mathbf{p}_{L.Toe})}{|\mathbf{p}_9 - \mathbf{p}_{L.Toe}|} \quad (B.40)$$

$$\mathbf{k}_6 = \frac{(\mathbf{p}_{L.Ankle} - \mathbf{p}_9) \times (\mathbf{p}_{L.Toe} - \mathbf{p}_9)}{|(\mathbf{p}_{L.Ankle} - \mathbf{p}_9) \times (\mathbf{p}_{L.Toe} - \mathbf{p}_9)|} \quad (B.41)$$

$$\mathbf{j}_6 = \mathbf{k}_6 \times \mathbf{i}_6 \quad (B.42)$$

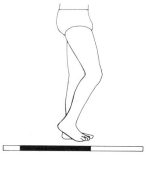

Frame = 22
Time = 0.84 s
Right midswing

It is important to realize that although these **ijk** vector triads are used to define the segmental coordinate system xyz, they are actually expressed in terms of the global reference system XYZ. The XYZ coordinates for the **ijk** vector triad of the pelvis and the six lower extremity segments are listed for time = 0.00 s in Table B.6 (which contains the data for the Man.REF file used in *GaitLab*.

Centers of Gravity

This section has three purposes: First, we provide the equations that are used to estimate centers of gravity based on joint centers and segment endpoints; second, we discuss the digital filter that is used to smooth raw position data; and third, we cover the finite difference theory that is the basis for performing numerical differentiation to calculate velocities and accelerations.

From Figure 3.11 and Tables 3.5 and B.2, the following equations may be derived:

$$\mathbf{p}_{R.Thigh.CG} = \mathbf{p}_{R.Hip} + 0.39\,(\mathbf{p}_{R.Knee} - \mathbf{p}_{R.Hip}) \quad (B.43)$$

$$\mathbf{p}_{L.Thigh.CG} = \mathbf{p}_{L.Hip} + 0.39\,(\mathbf{p}_{L.Knee} - \mathbf{p}_{L.Hip}) \quad (B.44)$$

$$\mathbf{p}_{R.Calf.CG} = \mathbf{p}_{R.Knee} + 0.42\,(\mathbf{p}_{R.Ankle} - \mathbf{p}_{R.Knee}) \quad (B.45)$$

$$\mathbf{p}_{L.Calf.CG} = \mathbf{p}_{L.Knee} + 0.42\,(\mathbf{p}_{L.Ankle} - \mathbf{p}_{L.Knee}) \quad (B.46)$$

$$\mathbf{p}_{R.Foot.CG} = \mathbf{p}_{R.Heel} + 0.44\,(\mathbf{p}_{R.Toe} - \mathbf{p}_{R.Heel}) \quad (B.47)$$

$$\mathbf{p}_{L.Foot.CG} = \mathbf{p}_{L.Heel} + 0.44\,(\mathbf{p}_{L.Toe} - \mathbf{p}_{L.Heel}) \quad (B.48)$$

In human movement activities such as gait, the frequency of the displacement signal is almost always less than the frequency of the noise. The purpose of a digital filter, therefore, is to filter out the high-frequency noise while allowing the low-frequency displacement signal to pass through untouched. The format of a low-pass digital filter is as follows:

$$x'_n = a_0 x_n + a_1 x_{n-1} + a_2 x_{n-2} + b_1 x'_{n-1} + b_2 x'_{n-2} \quad (B.49)$$

where x' refers to filtered output coordinates, x refers to raw unfiltered coordinate data, n refers to the nth sample frame, and a_0 through b_2 are the filter coefficients. These filter coefficients are constants that depend on the type and order of the filter, the sampling frequency (i.e., the frame rate), and the cutoff frequency (i.e., how much noise should be attenuated). As can be seen from Equation B.49, the filtered output x'_n is a weighted version of the immediate and past raw data, plus a weighted contribution of past filtered output. For the Gaitmath program in *GaitLab*, the second-order low-pass Butterworth filter was used. Further details may be obtained in Radar and Gold (1967) or Winter (1979). A FORTRAN listing of the subroutine DIGFIL which implements Equation A.49 may be found in Vaughan (1982).

We pointed out in chapter 3 that the digital filter has endpoint problems, which can lead to erroneous velocities and accelerations in the first few and last few frames. One of the algorithms that does not have these endpoint problems is the quintic spline

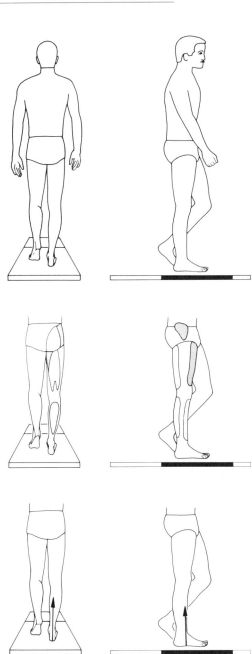

Frame = 8
Time = 0.28 s
Right midstance

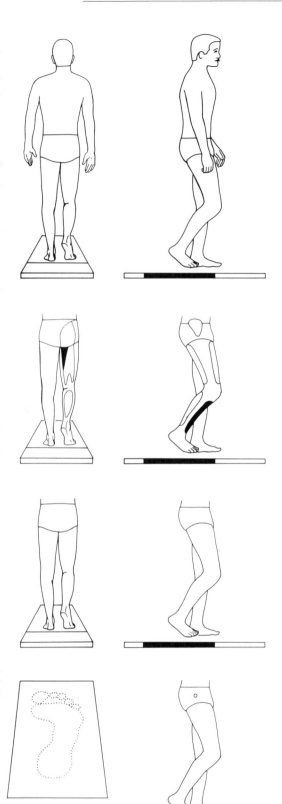

Table B.6 Three-Dimensional Coordinates of the ijk Unit Vectors for Segment Reference Frames at Time = 0.00 s (Right Heel Strike) for a Normal Male

Segment	i_X	i_Y	i_Z
Pelvis	−0.072	0.017	0.997
R. Thigh	−0.385	−0.088	0.919
L. Thigh	0.305	−0.001	0.952
R. Calf	−0.213	−0.016	0.977
L. Calf	0.606	0.001	0.795
R. Foot	−0.936	−0.095	−0.339
L. Foot	−0.884	0.043	0.465

Segment	j_X	j_Y	j_Z
Pelvis	0.996	0.053	0.071
R. Thigh	0.922	−0.008	0.386
L. Thigh	0.917	0.272	−0.293
R. Calf	0.977	−0.022	0.212
L. Calf	0.791	0.103	−0.603
R. Foot	−0.298	−0.302	0.906
L. Foot	0.454	0.309	0.835

Segment	k_X	k_Y	k_Z
Pelvis	−0.052	0.998	−0.021
R. Thigh	−0.026	0.996	0.084
L. Thigh	−0.258	0.962	0.084
R. Calf	0.018	1.000	0.021
L. Calf	−0.083	0.995	0.062
R. Foot	−0.188	0.949	0.254
L. Foot	−0.108	0.950	−0.294

Note. The XYZ values refer to the global coordinate system defined in Figure 3.10.

Frame = 21
Time = 0.80 s

(Vaughan, 1982; Wood & Jennings, 1979). We had planned to offer the quintic spline as an option for smoothing and differentiating in the Gaitmath software, but the size of the code and its running time precluded this option.

We have based our method for determining numerical differentiation on finite difference theory. Finite difference methods may be derived from Taylor series expansions (Miller & Nelson, 1973), and they provide formulae for calculating first and second derivatives of displacement-time data. The first and second derivatives (i.e., velocity and acceleration) are expressible as

$$\frac{dx_n}{dt} = \dot{x}_n = \frac{x_{n+1} - x_{n-1}}{2\Delta t} \quad \text{(B.50)}$$

and

$$\frac{d^2x_n}{dt^2} = \ddot{x}_n = \frac{x_{n+1} - 2x_n + x_{n-1}}{(\Delta t)^2} \quad \text{(B.51)}$$

where x is an input data point, n refers to the nth sample frame, and Δt is the time between adjacent frames. Equations B.50 and B.51 are known as central difference formulae. Forward and backward difference formulae may be used for derivatives of displacement data at the beginning and end of the data set. All these formulae are approximations, because the time interval Δt is not infinitely small. Therefore, any noise in the input signal has a large influence on the accuracy of the derivative values. A FORTRAN listing of the subroutine FIDIFF which implements Equations B.50 and B.51 is also included in the paper by Vaughan (1982).

Angular Kinematics

In this section we will cover three areas: definition of anatomical joint angles, definition of segment Euler angles, and derivation of segment angular velocities and accelerations based on the Euler angles.

We stated in chapter 3 that we chose to adopt the methods proposed by Chao (1980) and Grood and Suntay (1983) for defining our anatomical joint angles. Consider the segment reference frames defined in Figure 3.10. The lower extremities have been partitioned into six pairs of segments in Figure B.1, a-f.

The following conventions apply to all six joints:

$\mathbf{k}_{\text{Proximal}}$ = flexion/extension axis.
$\mathbf{i}_{\text{Distal}}$ = internal/external rotation axis.
$\mathbf{l}_{\text{Joint}}$ = abduction/adduction axis.

$$\mathbf{l}_{\text{Joint}} = \frac{\mathbf{k}_{\text{Proximal}} \times \mathbf{i}_{\text{Distal}}}{|\mathbf{k}_{\text{Proximal}} \times \mathbf{i}_{\text{Distal}}|} \quad \text{(B.52)}$$

α = flexion/extension angle.
β = abduction/adduction angle.
γ = internal/external rotation angle.

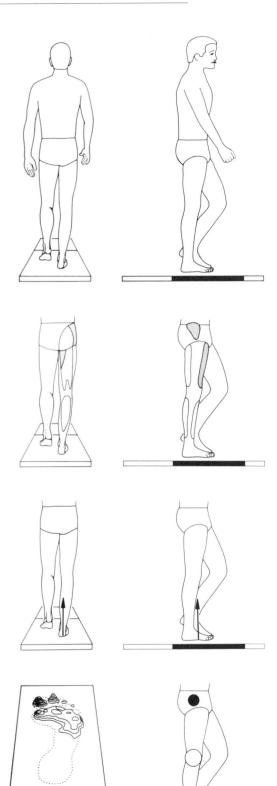

Frame = 9
Time = 0.32 s

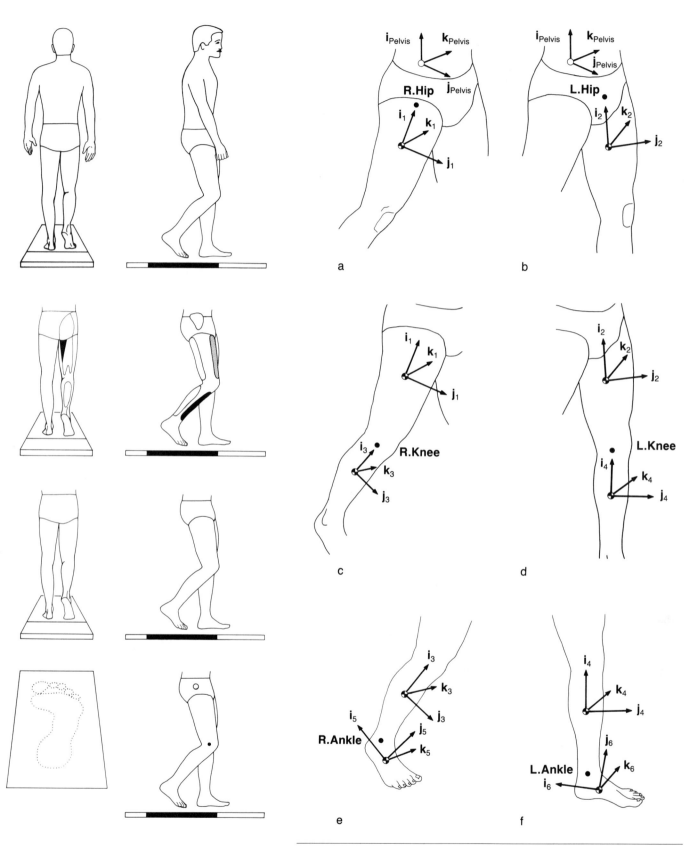

Figure B.1 Unit vector reference frames embedded in the proximal and distal segments on either side of an anatomical joint: (a) right hip; (b) left hip; (c) right knee; (d) left knee; (e) right ankle; (f) left ankle.

Flexion is positive and extension is negative.
Abduction is positive and adduction is negative.
Internal rotation is positive and external rotation is negative.

Using these conventions and the unit vector triads in Figure B.1, we get the following relationships for the anatomical joint angles:

$$\alpha_{R.Hip} = \sin^{-1}[\mathbf{l}_{R.Hip} \cdot \mathbf{i}_{Pelvis}] \quad (B.53)$$

$$\beta_{R.Hip} = \sin^{-1}[\mathbf{k}_{Pelvis} \cdot \mathbf{i}_1] \quad (B.54)$$

$$\gamma_{R.Hip} = -\sin^{-1}[\mathbf{l}_{R.Hip} \cdot \mathbf{k}_1] \quad (B.55)$$

$$\alpha_{L.Hip} = \sin^{-1}[\mathbf{l}_{L.Hip} \cdot \mathbf{i}_{Pelvis}] \quad (B.56)$$

$$\beta_{L.Hip} = -\sin^{-1}[\mathbf{k}_{Pelvis} \cdot \mathbf{i}_2] \quad (B.57)$$

$$\gamma_{L.Hip} = \sin^{-1}[\mathbf{l}_{L.Hip} \cdot \mathbf{k}_2] \quad (B.58)$$

$$\alpha_{R.Knee} = -\sin^{-1}[\mathbf{l}_{R.Knee} \cdot \mathbf{i}_1] \quad (B.59)$$

$$\beta_{R.Knee} = \sin^{-1}[\mathbf{k}_1 \cdot \mathbf{i}_3] \quad (B.60)$$

$$\gamma_{R.Knee} = -\sin^{-1}[\mathbf{l}_{R.Knee} \cdot \mathbf{k}_3] \quad (B.61)$$

$$\alpha_{L.Knee} = -\sin^{-1}[\mathbf{l}_{L.Knee} \cdot \mathbf{i}_2] \quad (B.62)$$

$$\beta_{L.Knee} = -\sin^{-1}[\mathbf{k}_2 \cdot \mathbf{i}_4] \quad (B.63)$$

$$\gamma_{L.Knee} = \sin^{-1}[\mathbf{l}_{L.Knee} \cdot \mathbf{k}_4] \quad (B.64)$$

$$\alpha_{R.Ankle} = \sin^{-1}[\mathbf{l}_{R.Ankle} \cdot \mathbf{j}_3] \quad (B.65)$$

$$\beta_{R.Ankle} = \sin^{-1}[\mathbf{k}_3 \cdot \mathbf{i}_5] \quad (B.66)$$

$$\gamma_{R.Ankle} = -\sin^{-1}[\mathbf{l}_{R.Ankle} \cdot \mathbf{k}_5] \quad (B.67)$$

$$\alpha_{L.Ankle} = \sin^{-1}[\mathbf{l}_{L.Ankle} \cdot \mathbf{j}_4] \quad (B.68)$$

$$\beta_{L.Ankle} = -\sin^{-1}[\mathbf{k}_4 \cdot \mathbf{i}_6] \quad (B.69)$$

$$\gamma_{L.Ankle} = \sin^{-1}[\mathbf{l}_{L.Ankle} \cdot \mathbf{k}_6] \quad (B.70)$$

Note that for the angles at the left and right ankle joints, the following conventions apply:

α = plantar flexion (positive) and dorsiflexion (negative)
β = varus (positive) and valgus (negative)
γ = inversion (positive) and eversion (negative)

The neutral position for determining plantar flexion and dorsiflexion is a right angle between the long axes of the calf and foot.

We showed in chapter 3 that a segment reference frame xyz may be oriented in 3-D space relative to the global reference system XYZ by means of three Euler angles. The Euler angle rotations are performed in the following order:

(a) ϕ about the **K** axis of the global reference frame,
(b) θ about the line of nodes, and
(c) ψ about the **k** axis of the segment,

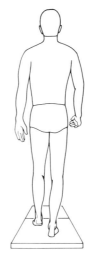
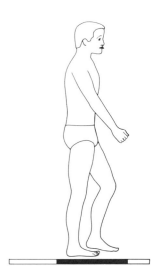

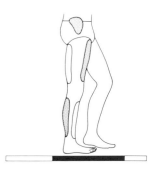

Frame = 10
Time = 0.36 s

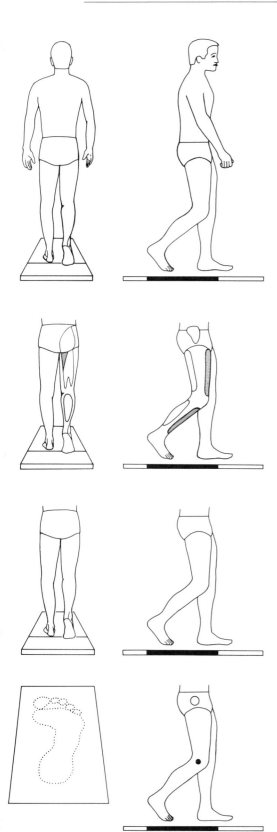

Frame = 19
Time = 0.72 s

where the line of nodes is a unit vector defined as

$$L = \frac{(K \times k)}{|K \times k|} \quad (B.71)$$

By way of example, Figure 3.16 has been expanded into Figure B.2, a-c, which shows each of the Euler angles for a single segment. The angles may be calculated as follows:

$$\phi = \sin^{-1}[(I \times L) \cdot K] \quad (B.72)$$

$$\theta = \sin^{-1}[(K \times k) \cdot L] \quad (B.73)$$

$$\psi = \sin^{-1}[(L \times i) \cdot k] \quad (B.74)$$

Our convention for the definition of the Euler angles is based on two classical mechanics texts by Synge and Griffith (1959) and Goldstein (1965). The segment angular velocities may be obtained from the Euler angles as follows:

$$\omega_{segment.x} = \dot{\phi}\sin\theta\sin\psi + \dot{\theta}\cos\psi \quad (B.75)$$

$$\omega_{segment.y} = \dot{\phi}\sin\theta\cos\psi - \dot{\theta}\sin\psi \quad (B.76)$$

$$\omega_{segment.z} = \dot{\phi}\cos\theta + \dot{\psi} \quad (B.77)$$

where the segment angular velocities ω are given relative to the segment-based reference frame xyz, and the dot above the Euler angles (e.g., $\dot{\phi}$) indicates the first derivative with respect to time (e.g., $\frac{d\phi}{dt}$). By taking the first derivative of Equations B.75 to B.77, we get the segment angular accelerations:

$$\dot{\omega}_{segment.x} = \ddot{\phi}\sin\theta\sin\psi + \dot{\phi}\dot{\theta}\cos\theta\sin\psi \quad (B.78)$$
$$+ \dot{\phi}\dot{\psi}\sin\theta\cos\psi + \ddot{\theta}\cos\psi - \dot{\theta}\dot{\psi}\sin\psi$$

$$\dot{\omega}_{segment.y} = \ddot{\phi}\sin\theta\cos\psi + \dot{\phi}\dot{\theta}\cos\theta\cos\psi \quad (B.79)$$
$$- \dot{\phi}\dot{\psi}\sin\theta\sin\psi - \ddot{\theta}\sin\psi - \dot{\theta}\dot{\psi}\cos\psi$$

$$\dot{\omega}_{segment.z} = \ddot{\phi}\cos\theta - \dot{\phi}\dot{\theta}\sin\theta + \ddot{\psi} \quad (B.80)$$

The Euler angles $\phi\theta\psi$ are smoothed using the digital filter described earlier in this chapter (Equation B.49), whereas finite difference methods (Equations B.50 and B.51) may be used to calculate first and second derivatives.

Equations B.75 to B.80 must be used to generate the angular velocities and accelerations for each of the six segments of the lower extremities (cf. Table 3.7), because these quantities are required to calculate the resultant joint moments described in the next section.

Dynamics of Joints

We are now at the stage where we can integrate all the previous sections and, using Newton's second and third laws of motion,

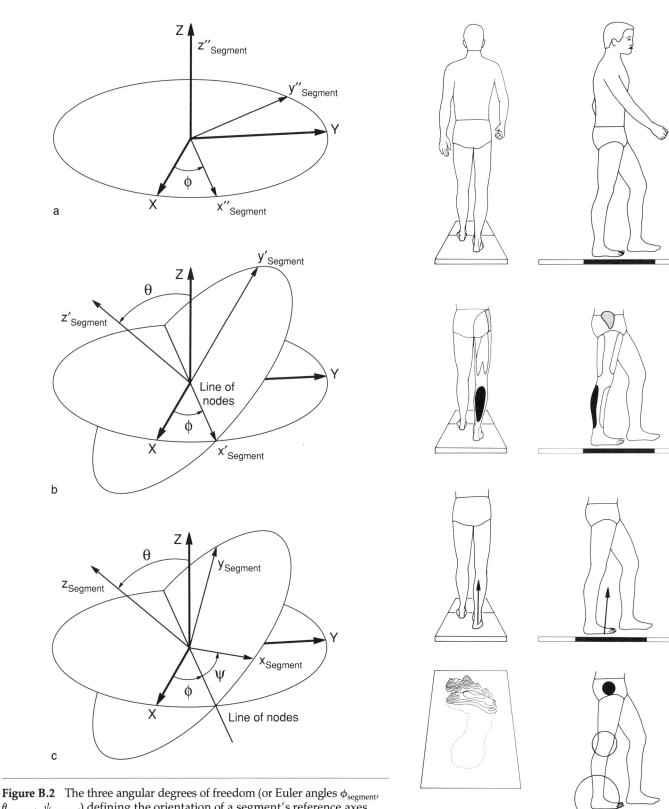

Figure B.2 The three angular degrees of freedom (or Euler angles $\phi_{segment}$, $\theta_{segment}$, $\psi_{segment}$) defining the orientation of a segment's reference axes ($x_{segment}$, $y_{segment}$, $z_{segment}$) relative to the global reference system XYZ (see Goldstein, 1965). Note that the CG has been moved to coincide with the origin of XYZ. The three Euler angle rotations take place in the following order: (a) $\phi_{segment}$ about the Z axis; (b) $\theta_{segment}$ about the line of nodes; and (c) $\psi_{segment}$ about the $z_{segment}$ axis. The line of nodes is perpendicular to both the Z and $z_{segment}$ axes. The primes and double primes indicate the intermediate axis positions.

Frame = 11
Time = 0.40 s

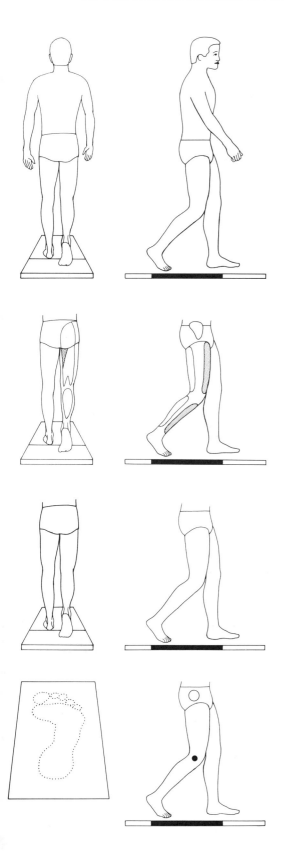

Frame = 18
Time = 0.68 s
Right toe-off

generate the resultant forces and moments acting at the lower extremity joints. In fact, we will integrate the following:

- Body segment parameters (BSP data file)
- Segment centers of gravity, velocities, and accelerations (COG data file)
- Ground reactions from force plates (FPL data file)
- Joint centers and segment endpoints (JNT data file)
- Segment reference frames (REF data file)
- Segment angular velocities and accelerations (ANG data file)

In performing this integration, we will follow a standard procedure of six steps for each of the segments:

1. Calculate the forces at the proximal joint using the linear form of Newton's second law.
2. Calculate the moment arms, proximal and distal, between the force application point and the segment center of gravity.
3. Calculate the residual moment acting on the segment.
4. Calculate the rate of change of angular momentum for the segment.
5. Calculate the resultant joint moment, first in the xyz system using the angular form of Newton's second law, then in the XYZ system.
6. Convert the joint force and moment from the XYZ system to a body-based system.

It is also pertinent to point out that these six steps are performed first on the foot, then on the calf, and finally on the thigh.

Because the format for the time rate of change of angular momentum is similar for all segments, angular momentum **H** and its first derivative **Ḣ** can be expressed in 3-D in terms of the segment reference frame:

$$\dot{\mathbf{H}}_{segment} = \dot{H}_{segment.x}\, \mathbf{i}_{segment} + \dot{H}_{segment.y}\, \mathbf{j}_{segment} + \dot{H}_{segment.z}\, \mathbf{k}_{segment} \quad \text{(B.81)}$$

The xyz components can be expressed in terms of moments of inertia, angular velocities, and angular accelerations (Goldstein, 1965):

$$\dot{H}_{segment.x} = I_{segment.IntExt}\, \dot{\omega}_{segment.x} + (I_{segment.FlxExt} - I_{segment.AbdAdd})\, \omega_{segment.z}\, \omega_{segment.y} \quad \text{(B.82)}$$

$$\dot{H}_{segment.y} = I_{segment.AbdAdd}\, \dot{\omega}_{segment.y} + (I_{segment.IntExt} - I_{segment.FlxExt})\, \omega_{segment.x}\, \omega_{segment.z} \quad \text{(B.83)}$$

$$\dot{H}_{segment.z} = I_{segment.FlxExt}\, \dot{\omega}_{segment.z} + (I_{segment.AbdAdd} - I_{segment.IntExt})\, \omega_{segment.y}\, \omega_{segment.x} \quad \text{(B.84)}$$

Right Foot. Application of the linear form of Newton's second law to the right foot yields the following:

$$F_{R.Ankle.X} = m_{R.Foot}\ddot{X}_{R.Foot.CG} - F_{Plate1.X} \quad (B.85)$$

$$F_{R.Ankle.Y} = m_{R.Foot}\ddot{Y}_{R.Foot.CG} - F_{Plate1.Y} \quad (B.86)$$

$$F_{R.Ankle.Z} = m_{R.Foot}(\ddot{Z}_{R.Foot.CG} - 9.81) - F_{Plate1.Z} \quad (B.87)$$

The proximal (Prx) and distal (Dis) moment arms may be calculated as follows:

$$\mathbf{p}_{Prx.5} = \mathbf{p}_{R.Ankle} - \mathbf{p}_{R.Foot.CG} \quad (B.88)$$

and

$$\mathbf{p}_{Dis.5} = \mathbf{p}_{Plate1} - \mathbf{p}_{R.Foot.CG} \quad (B.89)$$

where

$$\mathbf{p}_{Plate1} = DX1\mathbf{I} + DY1\mathbf{J} + 0\mathbf{K} \quad (B.90)$$

(The subscript 5 indicates the right foot.) The residual (Res) moment acting on the right foot is

$$\mathbf{M}_{Res.5} = \mathbf{T}_{Plate1} + (\mathbf{p}_{Prx.5} \times \mathbf{F}_{R.Ankle}) \quad (B.91)$$
$$+ (\mathbf{p}_{Dis.5} \times \mathbf{F}_{Plate1})$$

and

$$\mathbf{T}_{Plate1} = 0\mathbf{I} + 0\mathbf{J} + T_{Z.Plate1}\mathbf{K} \quad (B.92)$$

The rate of change of angular momentum for the right foot may be calculated using the standardized form of Equations B.82 to B.84. Then, application of the angular analog of Newton's second law yields

$$M_{R.Ankle.x} = \dot{H}_{5x} - \mathbf{i}_5 \cdot \mathbf{M}_{Res.5} \quad (B.93)$$

$$M_{R.Ankle.y} = \dot{H}_{5y} - \mathbf{j}_5 \cdot \mathbf{M}_{Res.5} \quad (B.94)$$

and

$$M_{R.Ankle.z} = \dot{H}_{5z} - \mathbf{k}_5 \cdot \mathbf{M}_{Res.5} \quad (B.95)$$

Adding these components gives

$$\mathbf{M}_{R.Ankle} = M_{R.Ankle.x}\mathbf{i}_5 + M_{R.Ankle.y}\mathbf{j}_5 + M_{R.Ankle.z}\mathbf{k}_5 \quad (B.96)$$

Because $\mathbf{i}_5\mathbf{j}_5\mathbf{k}_5$ are expressed in terms of the IJK (or XYZ) reference system (see Table B.6), Equation B.96 expresses $\mathbf{M}_{R.Ankle}$ also in terms of the XYZ system. From Equations B.85 to B.87,

$$\mathbf{F}_{R.Ankle} = F_{R.Ankle.X}\mathbf{I} + F_{R.Ankle.Y}\mathbf{J} + F_{R.Ankle.Z}\mathbf{K} \quad (B.97)$$

Equations B.97 and B.96 provide us with the resultant joint force (**F**) and moment (**M**) of the right calf acting on the right foot. These two vectors are expressed in terms of the global reference system XYZ. However, from an anatomical point of

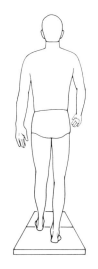
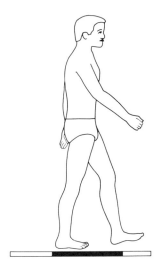

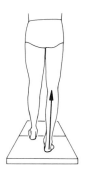
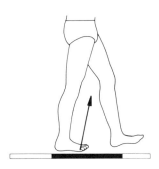

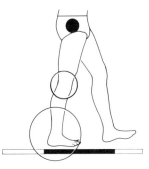

Frame = 12
Time = 0.44 s

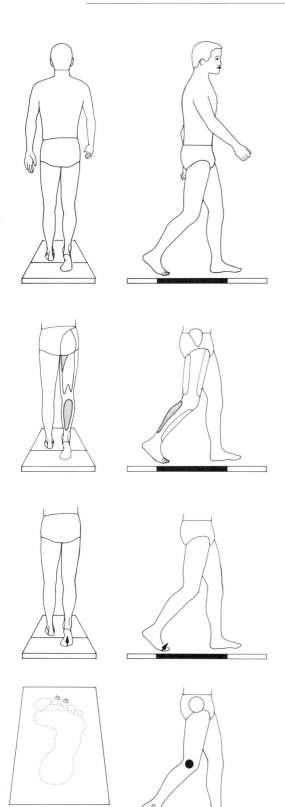

view, it makes far more sense to express the force and moment in terms of a body-based coordinate system. We have chosen to use the same coordinate systems that were used to calculate the anatomical joint angles (see Equations B.53 to B.70). Remember, too, that the resultant force or moment is being exerted by the proximal segment on the distal segment. Therefore, we get the following components:

$$F_{R.Ankle.PrxDis} = \mathbf{F}_{R.Ankle} \cdot \mathbf{i}_5 \quad (B.98)$$

$$F_{R.Ankle.MedLat} = \mathbf{F}_{R.Ankle} \cdot \mathbf{k}_3 \quad (B.99)$$

$$F_{R.Ankle.AntPos} = \mathbf{F}_{R.Ankle} \cdot \mathbf{l}_{R.Ankle} \quad (B.100)$$

Also,

$$M_{R.Ankle.InvEve} = \mathbf{M}_{R.Ankle} \cdot \mathbf{i}_5 \quad (B.101)$$

$$M_{R.Ankle.PlaDor} = \mathbf{M}_{R.Ankle} \cdot \mathbf{k}_3 \quad (B.102)$$

$$M_{R.Ankle.VarVal} = -\mathbf{M}_{R.Ankle} \cdot \mathbf{l}_{R.Ankle} \quad (B.103)$$

Right Calf. Application of the linear form of Newton's second law to the right calf yields the following:

$$F_{R.Knee.X} = m_{R.Calf}\ddot{X}_{R.Calf.CG} + F_{R.Ankle.X} \quad (B.104)$$

$$F_{R.Knee.Y} = m_{R.Calf}\ddot{Y}_{R.Calf.CG} + F_{R.Ankle.Y} \quad (B.105)$$

$$F_{R.Knee.Z} = m_{R.Calf}(\ddot{Z}_{R.Calf.CG} - 9.81) + F_{R.Ankle.Z} \quad (B.106)$$

The proximal (Prx) and distal (Dis) moment arms may be calculated as follows:

$$\mathbf{p}_{Prx.3} = \mathbf{p}_{R.Knee} - \mathbf{p}_{R.Calf.CG} \quad (B.107)$$

and

$$\mathbf{p}_{Dis.3} = \mathbf{p}_{R.Ankle} - \mathbf{p}_{R.Calf.CG} \quad (B.108)$$

where the subscript 3 refers to the right calf. The residual (Res) moment acting on the right calf is

$$\mathbf{M}_{Res.3} = -\mathbf{M}_{R.Ankle} - (\mathbf{p}_{Dis.3} \times \mathbf{F}_{R.Ankle}) \quad (B.109)$$
$$+ (\mathbf{p}_{Prx.3} \times \mathbf{F}_{R.Knee})$$

The rate of change of angular momentum for the right calf may be calculated using the standardized form of Equations B.82 to B.84. Then, application of the angular analog of Newton's second law yields the following:

$$M_{R.Knee.x} = \dot{H}_{3x} - \mathbf{i}_3 \cdot \mathbf{M}_{Res.3} \quad (B.110)$$

$$M_{R.Knee.y} = \dot{H}_{3y} - \mathbf{j}_3 \cdot \mathbf{M}_{Res.3} \quad (B.111)$$

$$M_{R.Knee.z} = \dot{H}_{3z} - \mathbf{k}_3 \cdot \mathbf{M}_{Res.3} \quad (B.112)$$

Adding these components gives

$$\mathbf{M}_{R.Knee} = M_{R.Knee.x}\mathbf{i}_3 + M_{R.Knee.y}\mathbf{j}_3 + M_{R.Knee.z}\mathbf{k}_3 \quad (B.113)$$

Again, we can express the resultant joint force (**F**; Equations B.104 to B.106) and resultant joint moment (**M**; Equation B.113) in terms of a body-based coordinate system:

$$F_{R.Knee.PrxDis} = \mathbf{F}_{R.Knee} \cdot \mathbf{i}_3 \quad (B.114)$$

$$F_{R.Knee.MedLat} = \mathbf{F}_{R.Knee} \cdot \mathbf{k}_1 \quad (B.115)$$

$$F_{R.Knee.AntPos} = \mathbf{F}_{R.Knee} \cdot \mathbf{l}_{R.Knee} \quad (B.116)$$

Also,

$$M_{R.Knee.IntExt} = \mathbf{M}_{R.Knee} \cdot \mathbf{i}_3 \quad (B.117)$$

$$M_{R.Knee.FlxExt} = \mathbf{M}_{R.Knee} \cdot \mathbf{k}_1 \quad (B.118)$$

$$M_{R.Knee.AbdAdd} = -\mathbf{M}_{R.Knee} \cdot \mathbf{l}_{R.Knee} \quad (B.119)$$

Right Thigh. Application of the linear form of Newton's second law to the right thigh yields the following:

$$F_{R.Hip.X} = m_{R.Thigh} \ddot{X}_{R.Thigh.CG} + F_{R.Knee.X} \quad (B.120)$$

$$F_{R.Hip.Y} = m_{R.Thigh} \ddot{Y}_{R.Thigh.CG} + F_{R.Knee.Y} \quad (B.121)$$

$$F_{R.Hip.Z} = m_{R.Thigh}(\ddot{Z}_{R.Thigh.CG} - 9.81) + F_{R.Knee.Z} \quad (B.122)$$

The proximal (Prx) and distal (Dis) moment arms may be calculated as follows:

$$\mathbf{p}_{Prx.1} = \mathbf{p}_{R.Hip} - \mathbf{p}_{R.Thigh.CG} \quad (B.123)$$

and

$$\mathbf{p}_{Dis.1} = \mathbf{p}_{R.Knee} - \mathbf{p}_{R.Thigh.CG} \quad (B.124)$$

where the subscript 1 refers to the right thigh. The residual (Res) moment acting on the right thigh is

$$\mathbf{M}_{Res.1} = -\mathbf{M}_{R.Knee} - (\mathbf{p}_{Dis.1} \times \mathbf{F}_{R.Knee}) + (\mathbf{p}_{Prx.1} \times \mathbf{F}_{R.Hip}) \quad (B.125)$$

The rate of change of angular momentum for the right thigh may be calculated using the standardized form of Equations B.82 to B.84. Then, application of the angular analog of Newton's second law yields the following:

$$M_{R.Hip.x} = \dot{H}_{1x} - \mathbf{i}_1 \cdot \mathbf{M}_{Res.1} \quad (B.126)$$

$$M_{R.Hip.y} = \dot{H}_{1y} - \mathbf{j}_1 \cdot \mathbf{M}_{Res.1} \quad (B.127)$$

$$M_{R.Hip.z} = \dot{H}_{1z} - \mathbf{k}_1 \cdot \mathbf{M}_{Res.1} \quad (B.128)$$

Adding these components gives

$$\mathbf{M}_{R.Hip} = M_{R.Hip.x}\mathbf{i}_1 + M_{R.Hip.y}\mathbf{j}_1 + M_{R.Hip.z}\mathbf{k}_1 \quad (B.129)$$

We can express the resultant joint force (**F**; Equations B.120 to B.122) and resultant joint moment (**M**; Equation B.129) in terms of a body-based coordinate system:

$$F_{R.Hip.PrxDis} = \mathbf{F}_{R.Hip} \cdot \mathbf{i}_1 \quad (B.130)$$

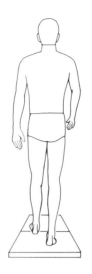
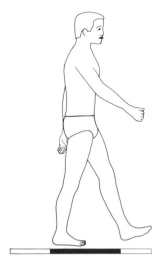
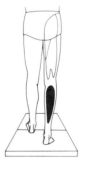

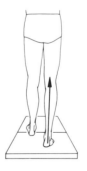
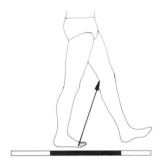

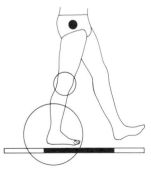

Frame = 13
Time = 0.48 s
Left heel strike

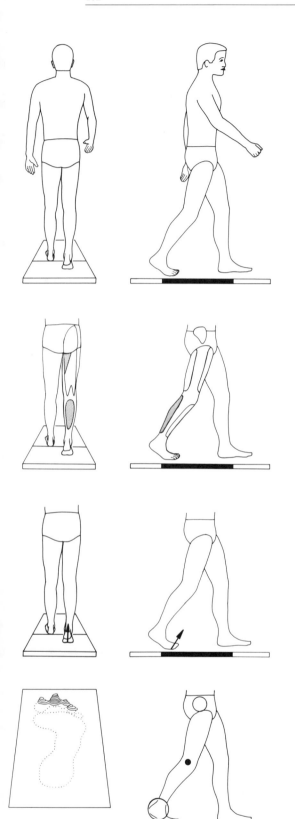

$$F_{R.Hip.MedLat} = \mathbf{F}_{R.Hip} \cdot \mathbf{k}_{Pelvis} \quad (B.131)$$

$$F_{R.Hip.AntPos} = \mathbf{F}_{R.Hip} \cdot \mathbf{l}_{R.Hip} \quad (B.132)$$

Also,

$$M_{R.Hip.IntExt} = \mathbf{M}_{R.Hip} \cdot \mathbf{i}_1 \quad (B.133)$$

$$M_{R.Hip.FlxExt} = -\mathbf{M}_{R.Hip} \cdot \mathbf{k}_{Pelvis} \quad (B.134)$$

$$M_{R.Hip.AbdAdd} = -\mathbf{M}_{R.Hip} \cdot \mathbf{l}_{R.Hip} \quad (B.135)$$

See the right leg free body diagrams in Figure B.3.

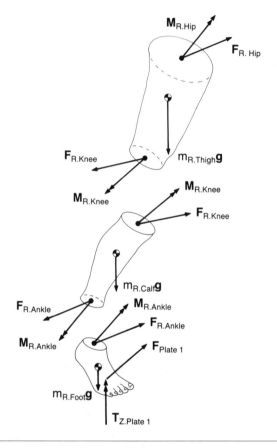

Figure B.3 Free body diagrams for the right foot, calf, and thigh, showing the external forces acting on each segment. Note that the forces and moments at the ankle and knee joints are equal in magnitude but opposite in direction, depending on the segment concerned (Newton's third law of motion).

Left Foot. Application of the linear form of Newton's second law to the left foot yields the following:

$$F_{L.Ankle.X} = m_{L.Foot} \ddot{X}_{L.Foot.CG} - F_{Plate2.X} \quad (B.136)$$

$$F_{L.Ankle.Y} = m_{L.Foot} \ddot{Y}_{L.Foot.CG} - F_{Plate2.Y} \quad (B.137)$$

$$F_{L.Ankle.Z} = m_{L.Foot} (\ddot{Z}_{L.Foot.CG} - 9.81) - F_{Plate2.Z} \quad (B.138)$$

The proximal (Prx) and distal (Dis) moment arms may be calculated as follows:

$$\mathbf{p}_{Prx.6} = \mathbf{p}_{L.Ankle} - \mathbf{p}_{L.Foot.CG} \qquad (B.139)$$

and

$$\mathbf{p}_{Dis.6} = \mathbf{p}_{Plate2} - \mathbf{p}_{L.Foot.CG} \qquad (B.140)$$

where

$$\mathbf{p}_{Plate2} = DX2\mathbf{I} + DY2\mathbf{J} + 0\mathbf{K} \qquad (B.141)$$

(The subscript 6 indicates the left foot.) The residual (Res) moment acting on the left foot is

$$\mathbf{M}_{Res.6} = \mathbf{T}_{Plate2} + (\mathbf{p}_{Prx.6} \times \mathbf{F}_{L.Ankle}) \qquad (B.142)$$
$$+ (\mathbf{p}_{Dis.6} \times \mathbf{F}_{Plate1})$$

and

$$\mathbf{T}_{Plate2} = 0\,\mathbf{I} + 0\,\mathbf{J} + T_{Z.Plate2}\,\mathbf{K} \qquad (B.143)$$

The rate of change of angular momentum for the left foot may be calculated using the standardized form of Equations B.82 to B.84. Then, application of the angular analog of Newton's second law yields

$$M_{L.Ankle.x} = \dot{H}_{6x} - \mathbf{i}_6 \cdot \mathbf{M}_{Res.6} \qquad (B.144)$$

$$M_{L.Ankle.y} = \dot{H}_{6y} - \mathbf{j}_6 \cdot \mathbf{M}_{Res.6} \qquad (B.145)$$

and

$$M_{L.Ankle.z} = \dot{H}_{6z} - \mathbf{k}_6 \cdot \mathbf{M}_{Res.6} \qquad (B.146)$$

Adding these components gives

$$\mathbf{M}_{L.Ankle} = M_{L.Ankle.x}\,\mathbf{i}_6 + M_{L.Ankle.y}\,\mathbf{j}_6 \qquad (B.147)$$
$$+ M_{L.Ankle.z}\,\mathbf{k}_6$$

Again, we can express the resultant joint force (**F**; Equations B.136 to B.138) and resultant joint moment (**M**; Equation B.147) in terms of a body-based coordinate system:

$$F_{L.Ankle.PrxDis} = \mathbf{F}_{L.Ankle} \cdot \mathbf{i}_6 \qquad (B.148)$$

$$F_{L.Ankle.MedLat} = -\mathbf{F}_{L.Ankle} \cdot \mathbf{k}_4 \qquad (B.149)$$

$$F_{L.Ankle.AntPos} = \mathbf{F}_{L.Ankle} \cdot \mathbf{l}_{L.Ankle} \qquad (B.150)$$

Also,

$$M_{L.Ankle.InvEve} = -\mathbf{M}_{L.Ankle} \cdot \mathbf{i}_6 \qquad (B.151)$$

$$M_{L.Ankle.PlaDor} = \mathbf{M}_{L.Ankle} \cdot \mathbf{k}_4 \qquad (B.152)$$

$$M_{L.Ankle.VarVal} = \mathbf{M}_{L.Ankle} \cdot \mathbf{l}_{L.Ankle} \qquad (B.153)$$

Left Calf. Application of the linear form of Newton's second law to the left calf yields the following:

$$F_{L.Knee.X} = m_{L.Calf}\ddot{X}_{L.Calf.CG} + F_{L.Ankle.X} \qquad (B.154)$$

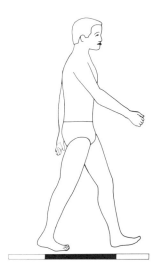

Frame = 14
Time = 0.52 s

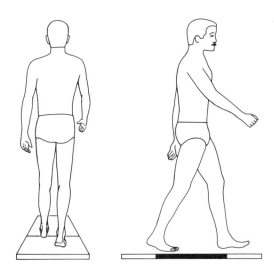

Frame = 15
Time = 0.56 s
Right heel-off

$$F_{L.Knee.Y} = m_{L.Calf} \ddot{Y}_{L.Calf.CG} + F_{L.Ankle.Y} \quad (B.155)$$

$$F_{L.Knee.Z} = m_{L.Calf} (\ddot{Z}_{L.Calf.CG} - 9.81) + F_{L.Ankle.Z} \quad (B.156)$$

The proximal (Prx) and distal (Dis) moment arms may be calculated as follows:

$$\mathbf{p}_{Prx.4} = \mathbf{p}_{L.Knee} - \mathbf{p}_{L.Calf.CG} \quad (B.157)$$

and

$$\mathbf{p}_{Dis.4} = \mathbf{p}_{L.Ankle} - \mathbf{p}_{L.Calf.CG} \quad (B.158)$$

where the subscript 4 refers to the left calf. The residual (Res) moment acting on the left calf is

$$\mathbf{M}_{Res.4} = -\mathbf{M}_{L.Ankle} - (\mathbf{p}_{Dis.4} \times \mathbf{F}_{L.Ankle}) \quad (B.159)$$
$$+ (\mathbf{p}_{Prx.4} \times \mathbf{F}_{L.Knee})$$

The rate of change of angular momentum for the left calf may be calculated using the standardized form of Equations B.82 to B.84. Then, application of the angular analog of Newton's second law yields the following:

$$M_{L.Knee.x} = \dot{H}_{4x} - \mathbf{i}_4 \cdot \mathbf{M}_{Res.4} \quad (B.160)$$

$$M_{L.Knee.y} = \dot{H}_{4y} - \mathbf{j}_4 \cdot \mathbf{M}_{Res.4} \quad (B.161)$$

$$M_{L.Knee.z} = \dot{H}_{4z} - \mathbf{k}_4 \cdot \mathbf{M}_{Res.4} \quad (B.162)$$

Adding these components gives

$$\mathbf{M}_{L.Knee} = M_{L.Knee.x} \mathbf{i}_4 + M_{L.Knee.y} \mathbf{j}_4 \quad (B.163)$$
$$+ M_{L.Knee.z} \mathbf{k}_4$$

We can express the resultant joint force (**F**; Equations B.154 to B.156) and resultant joint moment (**M**; Equation B.163) in terms of a body-based coordinate system:

$$F_{L.Knee.PrxDis} = \mathbf{F}_{L.Knee} \cdot \mathbf{i}_4 \quad (B.164)$$

$$F_{L.Knee.MedLat} = -\mathbf{F}_{L.Knee} \cdot \mathbf{k}_2 \quad (B.165)$$

$$F_{L.Knee.AntPos} = \mathbf{F}_{L.Knee} \cdot \mathbf{l}_{L.Knee} \quad (B.166)$$

Also,

$$M_{L.Knee.IntExt} = -\mathbf{M}_{L.Knee} \cdot \mathbf{i}_4 \quad (B.167)$$

$$M_{L.Knee.FlxExt} = \mathbf{M}_{L.Knee} \cdot \mathbf{k}_2 \quad (B.168)$$

$$M_{L.Knee.AbdAdd} = \mathbf{M}_{L.Knee} \cdot \mathbf{l}_{L.Knee} \quad (B.169)$$

Left Thigh. Application of the linear form of Newton's second law to the left thigh yields the following:

$$F_{L.Hip.X} = m_{L.Thigh} \ddot{X}_{L.Thigh.CG} + F_{L.Knee.X} \quad (B.170)$$

$$F_{L.Hip.Y} = m_{L.Thigh} \ddot{Y}_{L.Thigh.CG} + F_{L.Knee.Y} \quad (B.171)$$

$$F_{L.Hip.Z} = m_{L.Thigh} (\ddot{Z}_{L.Thigh.CG} - 9.81) + F_{L.Knee.Z} \quad (B.172)$$

The proximal (Prx) and distal (Dis) moment arms may be calculated as follows:

$$\mathbf{p}_{Prx.2} = \mathbf{p}_{L.Hip} - \mathbf{p}_{L.Thigh.CG} \quad (B.173)$$

and

$$\mathbf{p}_{Dis.2} = \mathbf{p}_{L.Knee} - \mathbf{p}_{L.Thigh.CG} \quad (B.174)$$

where the subscript 2 refers to the left thigh. The residual (Res) moment acting on the left thigh is

$$\mathbf{M}_{Res.2} = -\mathbf{M}_{L.Knee} - (\mathbf{p}_{Dis.2} \times \mathbf{F}_{L.Knee}) \quad (B.175)$$
$$+ (\mathbf{p}_{Prx.2} \times \mathbf{F}_{L.Hip})$$

The rate of change of angular momentum for the left thigh may be calculated using the standardized form of Equations B.82 to B.84. Then, application of the angular analog of Newton's second law yields the following:

$$M_{L.Hip.x} = \dot{H}_{2x} - \mathbf{i}_2 \cdot \mathbf{M}_{Res.2} \quad (B.176)$$

$$M_{L.Hip.y} = \dot{H}_{2y} - \mathbf{j}_2 \cdot \mathbf{M}_{Res.2} \quad (B.177)$$

$$M_{L.Hip.z} = \dot{H}_{2z} - \mathbf{k}_2 \cdot \mathbf{M}_{Res.2} \quad (B.178)$$

Adding these components gives

$$\mathbf{M}_{L.Hip} = M_{L.Hip.x}\,\mathbf{i}_2 + M_{L.Hip.y}\,\mathbf{j}_2 + M_{L.Hip.z}\,\mathbf{k}_2 \quad (B.179)$$

Again, we can express the resultant joint force (**F**; Equations B.170 to B.172) and resultant joint moment (**M**; Equation B.179) in terms of a body-based coordinate system:

$$F_{L.Hip.PrxDis} = \mathbf{F}_{L.Hip} \cdot \mathbf{i}_2 \quad (B.180)$$

$$F_{L.Hip.MedLat} = -\mathbf{F}_{L.Hip} \cdot \mathbf{k}_{Pelvis} \quad (B.181)$$

$$F_{L.Hip.AntPos} = \mathbf{F}_{L.Hip} \cdot \mathbf{l}_{L.Hip} \quad (B.182)$$

Also,

$$M_{L.Hip.IntExt} = -\mathbf{M}_{L.Hip} \cdot \mathbf{i}_2 \quad (B.183)$$

$$M_{L.Hip.FlxExt} = -\mathbf{M}_{L.Hip} \cdot \mathbf{k}_{Pelvis} \quad (B.184)$$

$$M_{L.Hip.AbdAdd} = \mathbf{M}_{L.Hip} \cdot \mathbf{l}_{L.Hip} \quad (B.185)$$

See the left leg free body diagrams in Figure B.4.

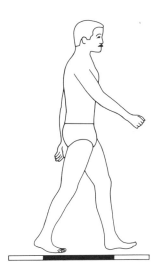

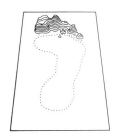

Frame = 15
Time = 0.56 s
Right heel-off

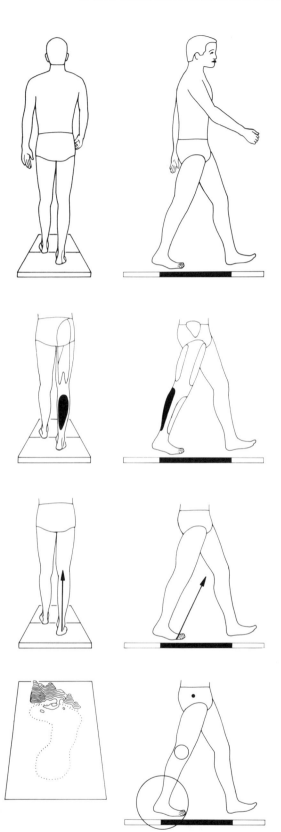

Frame = 14
Time = 0.52 s

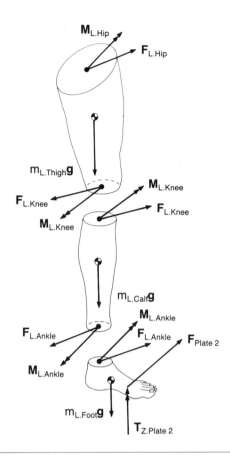

Figure B.4 Free body diagrams for the left foot, calf, and thigh, showing the external forces acting on each segment. Note that the forces and moments at the ankle and knee joints are equal in magnitude but opposite in direction, depending on the segment concerned (Newton's third law of motion).

APPENDIX C
Commercial Equipment for Gait Analysis

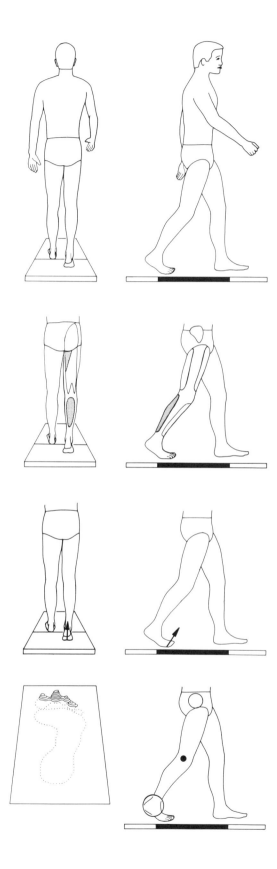

Frame = 16
Time = 0.60 s

In chapter 1 we introduced you to a framework for understanding the mechanics of human gait. The inverse dynamics approach of rigid body mechanics lets us make certain measurements and then use those data to say something about joint forces and moments and muscular tension. By way of more thickly outlined rectangular boxes, Figure 1.4 illustrates the four components in the movement chain that we can measure. These four components—electromyography, anthropometry, displacement of segments, and ground reaction forces—form the basis of this appendix. In addition, because we have stressed the importance of integration, there is a final section on software packages.

A number of companies have developed commercial products that are available for use by the gait analyst. We provide each company's name, address, telephone number, product names, and approximate price ranges. Our intent is not to recommend one product over another, but merely to describe the features of individual products and give you an up-to-date catalog of the instruments that are used in gait analysis. On the question of pricing we have given an approximate range, so that our information does not date too quickly and so that the companies have some flexibility. Bear in mind that all companies reserve the right to change their product specifications and prices without prior notice. The price ranges are as follows:

A Less than $5,000
B $5,000 to $10,000
C $10,000 to $20,000
D $20,000 to $30,000
E $30,000 to $50,000
F $50,000 to $100,000
G $100,000 to $150,000

107

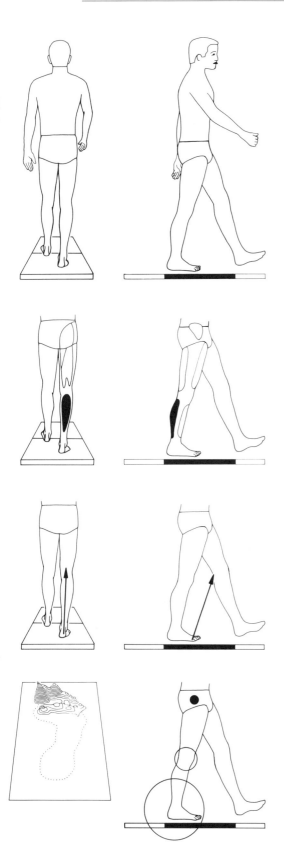

Frame = 13
Time = 0.48 s
Left heel-strike

Electromyography

There are many companies that produce electromyographic (EMG) equipment. Loeb and Gans (1986) have written an excellent book on EMG, including names and addresses of companies. If you would like to explore EMG techniques and equipment in more detail, refer to this book. In this section, we describe just two companies whose products have been developed primarily for the gait analyst.

Company Name: MIE Medical Research Limited
Address: 6 Wortley Moor Road
Leeds LS12 4JF
United Kingdom
Telephone: (0532) 793710
Product Name: MT8 Radio Telemetry System
Price Range: C

As the name suggests, this system frees the subject from being directly wired to the recording equipment. There are five main components: EMG preamplifiers; an 8-channel transmitter unit; a receiver unit; an analog-to-digital (A/D) card for connecting directly to an IBM PC (or compatible) microcomputer; and a software package. The preamplifiers, with a mass of 45 g including cable and connector and with a gain of 1,000 (*gain* is simply a multiplication factor), may be used either with the supplied electrode kit or with other commercially available, pregelled electrodes. The eight cables from the preamplifiers plug into the transmitter unit, which has a mass of 0.55 kg and is worn around the waist on a belt. The transmitter, powered by a rechargeable 9-V battery, has a line-of-sight range of greater than 150 m and may be used for applications other than EMG (cf. section on the displacement of segments). The receiver unit enables the gait analyst to choose appropriate gain, offset, and threshold settings for each individual channel as well as to test for the skin resistance. An added feature is its ability to recharge the transmitter unit's battery. The A/D card and software permit real-time viewing of the EMG signals, as well as the capture of data.

Company Name: Transkinetics
Address: 110 Shawmut Road
Canton, MA 02021
USA
Telephone: (617) 828-3000 or 1-800-441-6018
Product Name: System II and III Telemetry
Price Range: C

This product was designed originally for ambulatory electrocardiographic monitoring, but it has been successfully adapted for EMG analysis. Its strongest feature is the design of the electrode/preamplifier/transmitter, which is a single integrated unit with a mass of only 20 g. This eliminates the need for a backpack and means that the gait analyst need connect only as many channels as are required. The transmitter has a number of other attractive features: It can transmit signals from 30 Hz to 20 kHz, with signal levels from 30 µV peak-to-peak to 20 µV peak-to-peak, and can be used with either surface or wire electrodes. Line-of-sight range is about 70 m. The receiver module can accommodate up to 8 channels, and each of these can be low-pass or high-pass filtered. The maximum output voltage is 20 V peak-to-peak, and this signal can be readily connected to an oscilloscope or, via an A/D card, to a microcomputer.

Anthropometry

Although anthropometry may be broadly defined as the scientific measurement of the human body, in the context of gait analysis it simply means the measurement of certain features, such as total body mass or height, which enable the prediction of body segment parameters. These parameters are the segment masses and moments of inertia, the latter being a measure of the way in which the segment's mass is distributed about an axis of rotation. The simplest instruments required would be a bathroom scale and a flexible tape measure. Because such equipment is readily available (and can yield quite acceptable results), we will not review the whole field of companies that manufacture anthropometric equipment. As an example we will describe just one commercially available system.

Company Name: Carolina Biological Supply Company
Address: 2700 York Road
Burlington, NC 27215
USA
Telephone: 1-800-334-5551
Product Name: Anthropometric Measurements Set
Price Range: A

These research instruments are made of tool steel and have been specially designed for measuring stature and length of limb segments and portions of limbs. All the scales are in metric units and have vernier adjustments for added precision. There are special calipers for measuring such awkward parameters as joint diameter or chest depth. In addition, there are skinfold thickness calipers and precision flexible tape measures. The whole set comes in a durable, form-fitted storage case.

Displacement of Segments

There are many diseases of the neurological, muscular, and skeletal systems that manifest themselves as some form of movement dysfunction. It is not surprising, therefore, that many companies have concentrated on developing systems to measure the displacement of body segments. Two wide-ranging reviews on human movement were written by Atha (1984) and Woltring (1984), and you should refer to these papers for more detailed background information. Lanshammar (1985) has

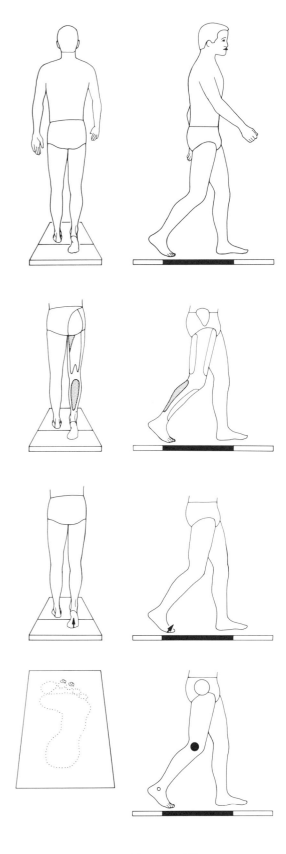

Frame = 17
Time = 0.64 s

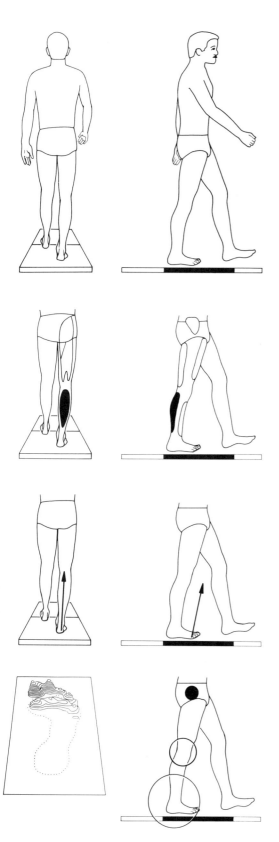

Frame = 12
Time = 0.44 s

suggested that the ideal device for the measurement and analysis of human displacement data would be characterized by

- high spatial resolution, better than 1:1,000;
- high sampling rate, at least 1,000 frames per second;
- passive, lightweight markers on the subject;
- automatic marker identification; and
- insensitivity to ambient light and reflections.

It comes as no surprise that there are no commercial systems currently available that meet all of the above criteria. The latest developments in this field have recently been published in the proceedings from an international meeting (Walton, 1990). These proceedings provide both a historical perspective and a fascinating insight to the field, showing just how close some companies are to realizing Lanshammar's goals. For the purposes of this review, we will consider the companies and their products grouped according to different technologies: electromechanical, ultrasound, and optical.

Company Name: MIE Medical Research Limited
Address: 6 Wortley Moor Road
Leeds LS12 4JF
United Kingdom
Telephone: (0532) 793710
Product Name: Gait Analysis System
Price Range: C

Like the electromyography equipment made by the same company, this system is based on telemetry and thus frees the subject or patient from being "hard-wired" to the recording instrument. There are six main components: toe and heel switches; electrogoniometers for the hip, knee, and ankle joints; an 8-channel transmitter unit; a receiver unit; an analog-to-digital card for connecting directly to an IBM PC (or compatible) microcomputer; and a software package for capturing and displaying the data. There are several advantages to this system: It is easy to operate; the data for a series of steps are available within minutes; other signals, such as heart rate, EMG, and foot pressure, can be transmitted simultaneously (bearing in mind the 8-channel limitation); no special laboratory facilities, other than the computer, are required. It has some disadvantages also: It encumbers the subject; the goniometers measure relative joint angles, rather than absolute joint positions; these particular goniometers are uniaxial, although, theoretically, multiaxial devices could be used.

Company Name: Chattecx Corporation
Address: PO Box 4287
Chattanooga, TN 37405
USA
Telephone: (615) 870-2281 or 1-800-322-7343
Product Name: Triax Three Dimensional Electro-Goniometer
Price Range: C

This system comes with the following five components: footswitches; triaxial goniometers for the left and right hip, knee, and ankle joints; analog-to-digital interface board; IBM PC compatible microcomputer; and appropriate driving software. The footswitches, which record heel and toe events, can be trimmed to fit inside the subject's shoe, or they can be worn with other "booties" under the sole. The goniometers are lightweight and come with variable length hardware to fit children as well as larger adults. The system has the following advantages: No special training is required to operate the equipment; the data are available within a matter of minutes; no special laboratory facilities, other than the computer, are required; and the angles measured at each joint are in three dimensions. Its disadvantages include the encumbrance of the subject, particularly because he or she must trail an umbilical cord of wires attached to the host computer, and the goniometers, which measure relative joint angles rather than absolute segment position.

Company Name: Penny & Giles Blackwood, Ltd.
Address: Blackwood Gwent
               NP2 2YD
               United Kingdom
Telephone: (0495) 228000
Product Name: Goniometer System, Angle Display Unit, and Data Recorder
Price Range: A

This unobtrusive and cost-effective system makes use of a relatively new technology—flexible strain gauges—to measure the orientation of one limb segment relative to another. These goniometers are very small (their masses range from 7 to 26 g) and far less bulky than the traditional electrogoniometers based on rotary potentiometers; these can be comfortably worn under normal clothing. The operating forces required to bend the transducer are negligible and will not restrict limb movement. Accurate alignment of the goniometer to the joint is not necessary (although alignment of each half of the transducer with the axis of the limb segment is desirable), and attachment can be accomplished simply with double-sided adhesive tape. The data are available in real-time, either on a hand-held, angle display unit, for input to an analog-to-digital converter card and computer (these latter two pieces of equipment are not part of the supplied system), or can be recorded on a multichannel data logger and replayed, stored, and analyzed by computer. These goniometers are sensitive to two axes of bending (e.g., flexion/extension with abduction/adduction), and they will soon have the third axis incorporated. Because of all the lead wires, this system still suffers from the problem of subject encumbrance, and the joint angles are relative rather than absolute, which makes integration with equipment that has a laboratory-based reference frame, such as force plates, extremely difficult. The adoption of the compact and powerful data logger permits the measurement of limb orientation during normal activity

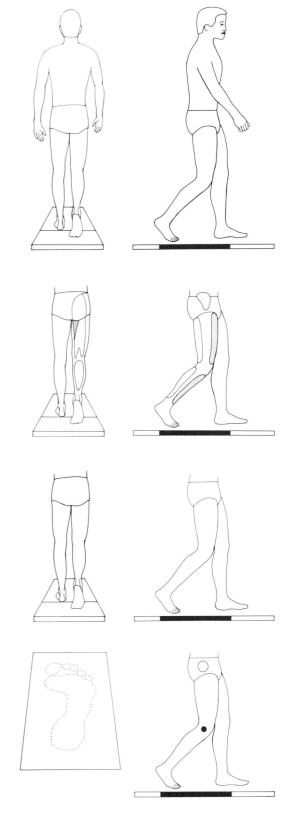

Frame = 18
Time = 0.68 s
Right toe-off

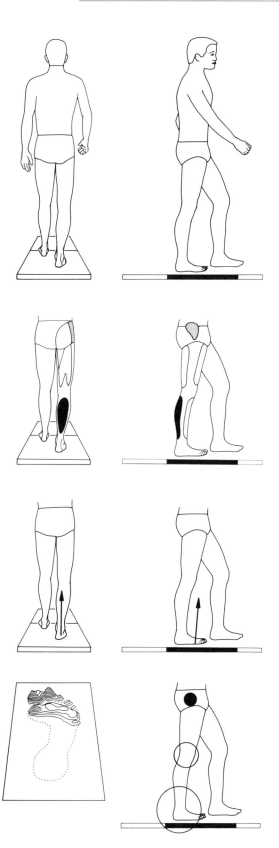

Frame = 11
Time = 0.40 s

without the constraints imposed by the clinical or laboratory environment.

Company Name:	Science Accessories Corporation
Address:	PO Box 550
	Southport, CT 06490
	USA
Telephone:	(203) 255-1526
Product Name:	GP-8-3D Digitizer
Price Range:	B

This three-dimensional digitizer is based on pulsed ultrasound (frequency 60 Hz). It has a user-adjustable active volume (up to 3.0 m × 3.0 m × 3.0 m), which is adequate for studying one or two strides in gait analysis. It consists of the following five components: up to eight separate emitters, which serve as markers; four microphones placed on a planar board; control electronics to activate a pulse from the emitters and measure the time for the pulse to reach each microphone; an RS-232 serial interface for an IBM PC or compatible computer (a parallel binary interface is optional); and software to capture and display the 3-D coordinates. The timing information for each pulse and each microphone can be converted into slant ranges, and in turn, Cartesian XYZ coordinates are calculated in software. However, the speed of sound (about 343 m/s in air) restricts the volume in which human movement can be monitored as well as the sampling rate. The sampling rate (about 10 Hz, when 8 markers are used) is limited, because the emitters have to be pulsed sequentially so they can be uniquely identified by the microphones. The spatial resolution is said to be 1:3,000, which is more than adequate for gait analysis, and the availability of data in real-time as well as unique marker identification are the GP-8-3D system's major advantages. Its disadvantages are a limited sampling rate; encumbrance of the subject by trailing wires; measurement limited to one side of the body, because the four receiving microphones must be placed on a plane; and sensitivity of the speed of sound to wind and temperature, although the latter can be corrected for in real-time by dedicating a single transmitter.

Company Name:	Instrumentation Marketing Corporation
Address:	820 South Mariposa Street
	Burbank, CA 91506
	USA
Telephone:	(213) 849-6251
Product Name:	High Speed Photographic & Videographic Motion Analysis Systems
Price Range:	B (Photo-Sonics 16-mm movie camera)
	C (NAC Film Analyzer)
	B (MOVIAS, Moving Image Analysis Software)

This equipment from IMC may be integrated to yield a gait analysis system that is based on two of the most mature technologies: movie cameras and XY digitizers. These provide very high sampling rates (up to 10,000 frames/s) and excellent spatial resolution (about 1:20,000). However, this level of performance comes with the following major disadvantages: Each marker position (e.g., knee joint) for each frame must be manually digitized, which is both tedious and time-consuming; movie film is expensive and takes time to be processed; three-dimensional data requires two or more cameras, increasing the data reduction time and costs considerably. The MOVIAS software, which is currently limited to XY (i.e., planar) data, runs on an IBM PC (or compatible) computer and allows the user to graph various kinematic and dynamic quantities.

Company Name:	Peak Performance Technologies, Inc.
Address:	7388 South Revere Parkway, Suite 801
	Englewood, CO 80112
	USA
Telephone:	(303) 799-8686
Product Name:	Peak Video Motion Measurement Systems
Price Range:	D to F

These are complete turn-key systems (i.e., everything is included, and all you need to do is "turn the key" to get the system up and running) consisting of the following three components: (1) a two-dimensional system, with video camcorder, video cassette recorder (VCR), video monitor, VCR controller board, video frame acquisition board, IBM PC/AT compatible computer with math coprocessor, graphics monitor, printer, and driving software; (2) a three-dimensional system, with additional video cameras that can be synchronized with the master camcorder, a portable VCR, a calibration frame, and appropriate 3-D module software; and (3) an automatic system, with flood lights, reflective markers, and additional software. The temporal resolution of Peak Systems is variable depending on the video recording system being used. The standard system arrangement uses 60 frames/s, although the Peak System is compatible with video recording equipment that can record at a rate of up to 200 frames/s. The spatial resolution is 1:500. The advantages of these systems are as follows: Markers are not always required; movement can be captured on videotape (even under adverse field and lighting conditions) and then processed by the computer at a later time; the software to process and display the kinematic information is very flexible, creating animated stick figures for qualitative assessment and parameter graphs (such as linear and angular displacement, velocity, and acceleration) for quantitative analysis. The major disadvantages are that systems (1) and (2) require considerable "hands-on" from the operator to digitize the data (although this can be partially alleviated by implementing system [3]), and the time from capturing the movement of interest to the availability of data can be quite lengthy.

Company Name:	Motion Analysis Corporation
Address:	3650 North Laughlin Road
	Santa Rosa, CA 95403
	USA
Telephone:	(707) 579-6511 or 1-800-862-1338
Product Name:	ExpertVision and FootTrak
Price Range:	E (2-D system)
	G (3-D system)

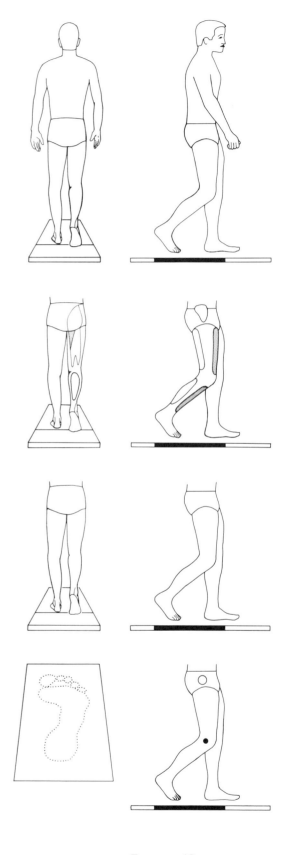

Frame = 19
Time = 0.72 s

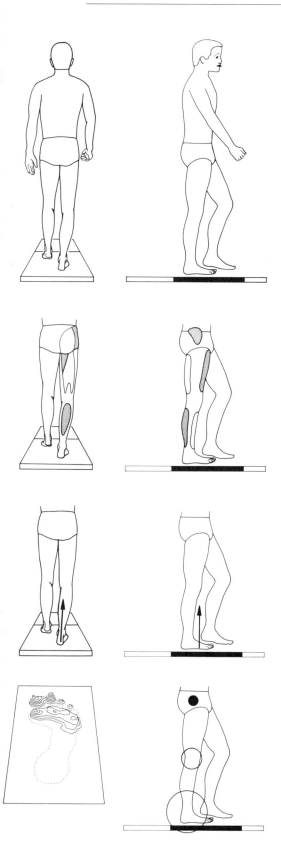

Frame = 10
Time = 0.36 s

Both the 2-D and 3-D systems include a computer: an IBM PC/386 and a Sun Graphics Workstation, respectively. Therefore, they are both so-called turn-key systems. The primary components of each system are retroreflective markers, video cameras, video recorders, a video monitor, a video image processor, the computer plus printer or plotter, and the driving software. The FootTrak system has been designed specifically for the needs of podiatrists and sports medicine clinics. It is a 2-D system and enables the user to study the calf and foot during the stance phase as the patient walks or runs on a treadmill. The standard data capture rate for the 3-D ExpertVision system is 60 frames/s, although 200 frames/s can be accomplished using special high-speed video cameras and videotaping. The spatial resolution is about 1:1,500. Although the generation of 3-D coordinates is automatic, a 15-marker gait trial can still take up to 30 min to process all the data. This time delay is a function of the marker-tracking algorithm and the speed of the computer. Major advantages of the system are its versatility: Movements can be studied indoors or outdoors; other relevant data, such as electromyography and ground reaction forces, can be gathered simultaneously and integrated with the kinematics; and up to six cameras can be used at one time. A software package called OrthoTrak, designed specifically for gait analysis, is also available and is described in the last section of this appendix.

Company Name:	Oxford Metrics Ltd.	
Address:	Unit 8, 7 West Way	14206 Carlson Circle
	Oxford OX2 0JB	Tampa, FL 33625
	England	USA
Telephone:	(0865) 244656	(813) 855-2910
Product Name:	VICON	
Price Range:	F	

The VICON System for gait analysis has the following basic components: retroreflective passive markers; video cameras, which operate at 50, 60, or 200 frames/s and incorporate solid state sensors and synchronous shutters; a proprietary video interface for digitizing the analog signal; a computer with graphics terminal (based on Digital Equipment Corporation's PDP/VAX series); and driving software. Depending on the camera used, the spatial resolution can be as high as 1:1,000. A recent hardware development, the Etherbox, is an intelligent stand-alone system for video coordinate generation and storage. With inputs from two to seven cameras (their placement is unrestricted), the Etherbox can determine the 3-D coordinates of more than 30 different markers. The Etherbox can also digitize up to 64 analog channels for capturing (at frequencies up to 5 kHz) EMG and ground reaction force data simultaneous to recording the video data; it communicates with the host computer via the Ethernet Local Area Network. The applications software has been modularized, enabling the end-user to configure and customize the graphic output to his or her particular needs. Probably the major disadvantage of the VICON System is the difficulty it has in tracking (i.e., identifying) individual markers when multiple cameras and more than 10 markers are used. This increases the processing time, because the user must identify each marker trajectory individually. However, this problem has been solved by an add-on software product from another company, ADTECH, which is described later in this chapter. The standard data file structure developed over the years by Oxford Metrics has enabled a company like ADTECH and other gait laboratories to write software that can be exchanged among VICON users.

Company Name:	Bioengineering Technology & Systems
Address:	via Capecelatro 66-I
	20148 Milano
	Italy
Telephone:	(02) 4043951 or 4047896
Product Name:	Elite Motion Analyser
Price Range:	E (excluding computer and peripherals)

The major components of this system are passive retroreflective markers (diameter from 1 mm to 1 cm); high sensitivity video cameras and either a visible or infrared light source; a video image processor that digitizes the analog signal and recognizes individual markers using a numerical "mask" algorithm; a computer with color graphics terminal (based on Digital Equipment Corporation's PDP series); and software to calibrate, capture, and display the data. Force platform and EMG data may be gathered simultaneously. The standard sampling rates are either 50 or 100 frames/s, and the system accuracy is claimed to be 1:2,500 (for 2-D) and 1:1,500 (for 3-D). Up to four separate cameras can be used with the video image processor; but, as with most video-based systems that use passive markers, the identification of the individual markers remains a problem that is not entirely handled by the software alone. Some user input is required.

Company Name:	Human Performance Technologies, Inc.
Address:	12 Technology Drive, Suite 6
	East Setauket, NY 11733
	USA
Telephone:	(516) 689-6521
Product Name:	2K-QDA
Price Range:	E (for 2-D and 3-D)

These systems make use of solid state video cameras (1 for 2-D; 2 for 3-D), a real-time image processor, reflective markers, video recorders, black-and-white monitors, optical light pen, and two IBM PC/XT compatible computers. EMG and pressure insole systems are also included, and the data may be gathered simultaneously with the video data. The 2-D video system can best be described as semiautomatic: The user must digitize all the marker positions with the light pen for the first few frames, and thereafter the software algorithm can track the marker positions automatically. The 3-D system requires that the two cameras be placed with their optical axes exactly orthogonal to each other. The sampling rate is 60 frames/s, and spatial resolution is approximately 1:500. The major advantage of these systems is that they are designed to capture force and muscle actions at the same time as the segment displacements. However, they suffer from the disadvantage of requiring considerable "hands-on"

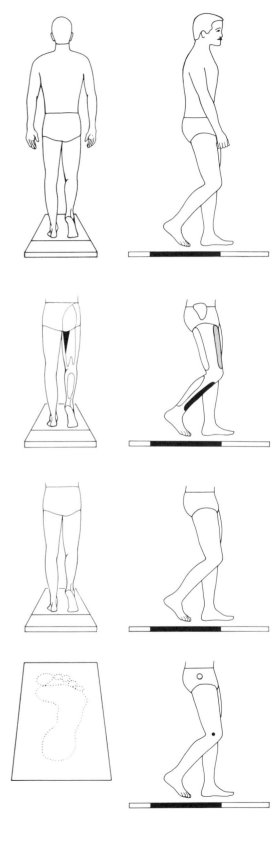

Frame = 20
Time = 0.76 s
Right acceleration

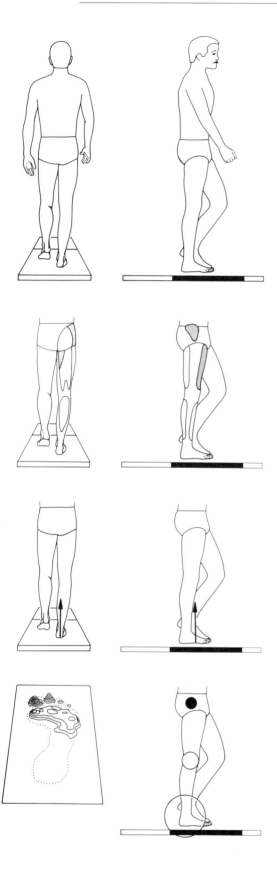

Frame = 9
Time = 0.32 s

from the user, and as with other video-based systems, marker identification is a problem in automatic mode.

Company Name: Columbus Instruments
Address: PO Box 44049
Columbus, OH 43204
USA
Telephone: (614) 488-6176
Product Name: Videomex-X
Price Range: C (2-D system)
D (3-D system)

The Videomex-X is comprised of the following components: colored markers (up to six) to be placed on the subject; an RGB color video camera (one for the 2-D system; two for the 3-D); a high-speed color image analyzer; an IBM AT compatible computer; and motion analysis software. It can combine passive markers (which do not encumber the subject) and unique marker identification in real-time by color-coding. Sampling rate is either 30 or 60 frames/s, and with the pixel resolution of 240 × 240, the system's spatial resolution is approximately 1:500. The major advantage of these systems is the availability of the data in real-time, without the need for tedious postprocessing to track and identify markers. The disadvantages include the limited number of markers (six), the large size of markers (at least 2% of field-of-view width), and only two cameras for 3-D, which limits you to studying only one side of the body at a time.

Company Name: Loredan
Address: PO Box 1154
Davis, CA 95617
USA
Telephone: (916) 758-3622
Product Name: LIDO Kinetic Analysis System
Price Range: E

The LIDO system consists of passive retroreflective markers; a black-and-white CCD camera; a VCR with freeze-field capability for 60 Hz resolution; a video character generator for adding titles to video images; a video signal processor for automatic marker digitization; a black-and-white video monitor; and an IBM PS/2 model 30 microcomputer with VGA graphics. This 2-D system has a sampling rate of 60 Hz, spatial resolution of approximately 1:500, and the option of a 16-channel A/D converter for capturing analog signals, such as electromyography and ground reaction forces. This system suffers from the disadvantage of requiring considerable "hands-on" from the user, and with more than six markers, identification in automatic mode is a major problem.

Company Name: Motiontronics for Science
Address: 1118 East 215th Street
Bronx, NY 10469
USA
Telephone: (212) 798-7497
Product Name: Motion Analysis System Engine
Price Range: A

This system comprises just three components: retroreflective markers; a plug-in video digitizer board; and software to capture 2-D data and perform biomechanical analysis of walking and running. The user must supply a standard black-and-white video camera, and an IBM AT compatible microcomputer. Its major advantage is the availability of the XY coordinates in real-time and the simplicity of the system. However, it can only capture 2-D data, and as with most other video-based systems, marker identification of a large number of markers becomes a problem.

Company Name:	Selective Electronics, Inc.	
Address:	21654 Melrose	Flojelbergsgatan 14
	Southfield, MI 48075	S-431 37 Molndal
	USA	Sweden
Telephone:	(313) 355-5900	46-31-878110
Product Name:	Selspot II Cameras and MULTILab	
Price Range:	E to F	

Selspot has been marketing systems to measure the displacement of segments for over a decade. Their system is based on active markers (infrared-emitting diodes) that are pulsed sequentially and a camera, which contains a dual-axis photodiode that records the "center of gravity" of the incoming light signal. This system provides a very high sampling rate (over 600 frames/s for 15 markers), excellent spatial resolution (1:4,000), and unique marker identification. It has two disadvantages: (1) Because the markers are active, they must be connected to the driving electronics, either by an umbilical cord or by telemetry, thus adding to the encumbrance of the subject; and (2) because the photodiode looks at the center of gravity, reflections from shiny surfaces (such as the floor) lead to two separate light signals and thus erroneous position data (software filters that reduce the effects of reflections are supplied in the MULTILab package). The Selspot II cameras have recently been integrated into MULTILab, which is a stand-alone measurement and data-processing system. It has been set up so the user can configure his or her own system. The three primary components are a computer processor based on the standard VME bus or an IBM PC/AT compatible; software that facilitates data collection, processing (a mathematical calculator with powerful commands), and presentation; and sensor systems such as the Selspot II cameras, force platforms, and EMG amplifiers.

Company Name:	Northern Digital, Inc.
Address:	403 Albert Street
	Waterloo, ON N2L 3V2
	Canada
Telephone:	(519) 884-5142
Product Name:	WATSMART
Price Range:	E (Basic system)
	E (Basic system plus real-time card and 16-channel analog acquisition unit)

The major components of the WATSMART system include 24 active, infrared light-emitting-diode markers; three strobe units that aid in exciting the markers sequentially (in groups of 8); a

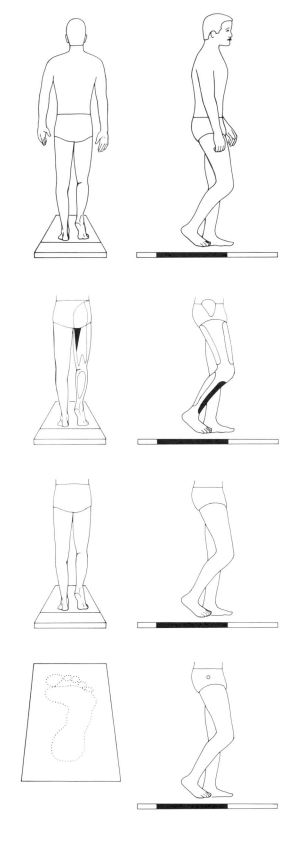

Frame = 21
Time = 0.80 s

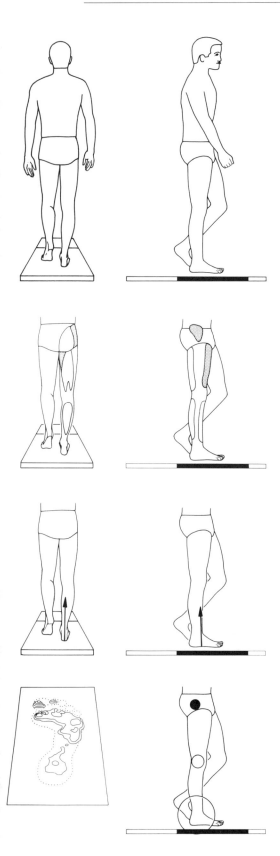

Frame = 8
Time = 0.28 s
Right midstance

system unit; two optoelectronic cameras based on lateral effect photodiodes; a calibration frame (available in three sizes); and software for data collection, 3-D reconstruction, and graphics display that runs on an IBM PC or compatible computer. Options include a real-time card that collects and reconstructs 4,000 3-D points/s (i.e., 200 frames/s for a 20-marker configuration) and the WATSCOPE data acquisition unit that provides 16 analog, input channels. The absolute accuracy of the system in 3-D is 1:500 across the 33° field of view for each axis, although each camera can distinguish 4,000 positions across that field. Advantages of the system include high sampling rate, wide viewing angle, good spatial accuracy and resolution, unique marker identification, 3-D data available in real-time, and that it runs on a standard IBM PC or compatible computer. Its disadvantages are the encumbrance of the subject by the trailing umbilical wires which are required to drive each individual marker, and the possibility of reflections which adversely affect the accuracy of the two-dimensional coordinates of a marker. A recent introduction from Northern Digital is the OPTOTRAK system, based on multiple linear CCD (2048) technology, which has significantly better accuracy and spatial resolution than WATSMART. However, it still uses active markers, so there are encumbrance problems.

Ground Reaction Forces

Our interest in the forces and pressures acting on the soles of our feet is by no means new. Over a century ago, Marey (1886) developed one of the earliest systems to measure ground reaction forces. A fixed force plate, developed by Fenn (1930) and designed to measure forces in three orthogonal directions, has been in existence for over half a century. Today there are two essential types of commercial devices for measuring ground reaction forces: force plates, which are fixed in the ground and record the force between the ground and the plantar surface of the foot (or sole, if the subject is wearing shoes); and pressure insoles, which are worn inside the shoe and record the pressures between the plantar surface of the foot and the shoe sole. The force plate is stationary and can only record the stance phase of a single gait cycle, whereas the pressure insole moves with the subject and can record multiple steps.

Company Name:	Kistler Instrument Corporation	
Address:	75 John Glenn Drive	Eulachstrasse 22
	Amherst, NY 14120	CH-8408 Winterthur
	USA	Switzerland
Telephone:	(716) 691-5100	(052) 831111
Product Name:	Kistler Force Plate	
Price Range:	C	

Kistler Force Plates have been commercially available for over 2 decades. They are based on piezoelectric quartz transducers that are sensitive to loads in the three orthogonal directions. One of these triaxial transducers is mounted near each of the four corners of the plate, so that the device provides the following information: reaction forces in the X, Y, and Z directions; the X, Y position of this resultant reaction force; and the free moment about the vertical (Z) axis, as well as the moments about the X and Y axes. Because of the natural tendency of piezoelectric materials to provide a decaying signal when placed under a static load, special charge amplifiers, which provide a drift rate of 0.03 pC/s, are supplied with the system. This combination of transducer and amplifier provides a wide operating range ($-10,000$ to $+20,000$ N) and yet is very sensitive (it can measure the heart rate of a person standing quietly on the plate). Depending on the plate's material (steel, aluminum, or glass) and its size, the natural frequency varies from 300 to over 1,000 Hz, which is quite adequate for gait analysis. The standard size is 0.6 m × 0.4 m × 0.1 m, although larger sizes can be obtained. The major advantage of the Kistler Force Plate is that it provides all the ground reaction force information necessary for doing a dynamic analysis of lower extremity gait. However, its disadvantage is that though it provides the resultant ground reaction force and its point of application, it provides no information on the distribution of this force (i.e., the pressure).

Company Name:	Advanced Mechanical Technology, Inc.
Address:	151 California Street
	Newton, MA 02158
	USA
Telephone:	(617) 964-2042
Product Name:	BIOVEC-1000 Force Platform System
Price Range:	D

The biomechanics force platform systems from AMTI are based on strain gauges, a mature and well-established force-transducing technology. In addition to the force plate itself, the system comes with a multichannel strain gauge amplifier and integrated software that runs on IBM PC and compatible computers. The plate provides six outputs—the forces and moments about the orthogonal X, Y, and Z axes. The standard plate has the following features: a loading range of 2,500 to 5,000 N, resonant frequency of 250 to 500 Hz, and a size of 0.51 m × 0.46 m × 0.08 m. Other plates with larger loading capacities, higher resonant frequencies, or larger areas are also available. The major advantage of the AMTI force plate is that it provides all the ground reaction force information necessary for doing a dynamic analysis of lower extremity gait. However, its disadvantage is that it provides the resultant ground reaction force and its point of application, but provides no information on the distribution of this force (i.e., the pressure).

Company Name:	Baltimore Therapeutic Equipment Co.
Address:	7455-L New Ridge Road
	Hanover, MD 21076
	USA
Telephone:	(301) 850-0333
Product Name:	Dynamic Pedobarograph
Price Range:	E

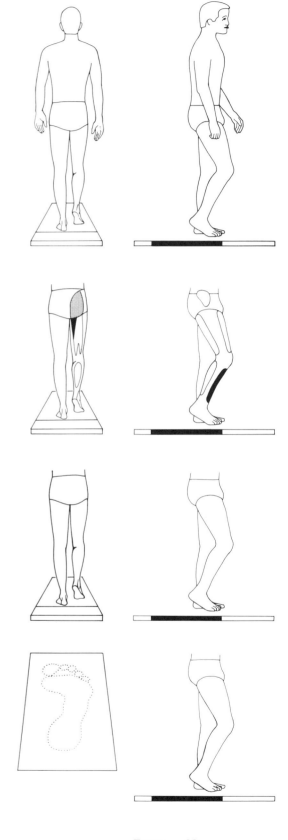

Frame = 22
Time = 0.84 s
Right midswing

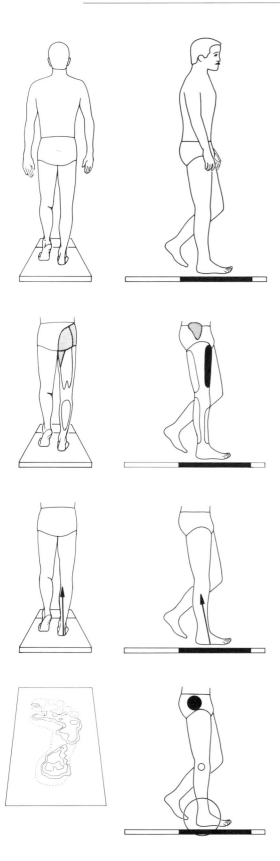

Frame = 7
Time = 0.24 s

The Dynamic Pedobarograph measures pressure by viewing, with a sensitive video camera, the underside of a lighted glass plate upon which the subject walks. The pressure of the footstep causes light to escape from the glass. The amount of light escaping varies according to the pressure at each point on the plate. A high-speed microcomputer (IBM PC/AT or compatible) connected to a special digital interface converts the image from the video camera into a digital signal. The system can capture 30 images/s in real-time and resolve pressure areas as small as 2 mm × 3 mm. Each corner of the plate is equipped with strain gauges so that the resultant vertical force acting on the plate can be recorded simultaneously for calibration purposes. The plate area is 0.38 m × 0.28 m, the pressure resolution is 4 kPa, and the pressure range is 1.1 MPa. The plantar pressure is displayed on the computer monitor as a color-coded contour diagram, which very clearly highlights the areas of increased pressure (e.g., over the metatarsal heads). These areas can be studied in more detail by placing a circular mask over the area of interest with the aid of a mouse. The advantages of this system are its simplicity of operation, intuitive method of displaying pressure data, excellent spatial and pressure resolution, and adequate pressure range. The disadvantages are a relatively low sampling rate, the small size of platform, which could lead to targeting, its lack of portability and need of installation in a special walkway, and the lack of shear force information (i.e., no anterior/posterior or medial/lateral forces).

Company Name: Polysens SpA
Address: via Donatello 24
50028 Tavarnelle Val di Pesa
Florence
Italy
Telephone: (3955) 8071580
Product Name: Orthomat
Price Range: D

The Orthomat measures pressure between the bottom of the foot (or shoe) and the ground using a matrix of piezoelectric (polyvinylidene fluoride) sensors. The area of the platform is 32 cm × 32 cm, and there are 1,024 sensors, thus providing one sensor/cm^2. The minimum pressure detectable on each sensor is 400 Pa; the maximum is 1.5 MPa. The sampling frequency may be set between 12.5 and 100 Hz and a maximum of 512 frames of data can be captured. The Orthomat connects directly to a standard IBM XT or AT and requires an EGA graphics card and compatible color monitor (the computer is not included with the system). The software is very flexible and allows the user to extract the following parameters: contact area between foot and ground, spatial distribution of load, center of pressure, and maximum load. It also provides 3-D animation, viewing the pressure profiles from different perspectives. As with other similar systems, the pressure levels are color-coded for ease of understanding. The advantages of this system are its simplicity of operation, the intuitive method of displaying pressure data, good spatial and pressure resolution, and its portability (the platform has a mass of only 15 kg). The disadvantages are the small platform size, which could lead to targeting, and the lack of shear force information.

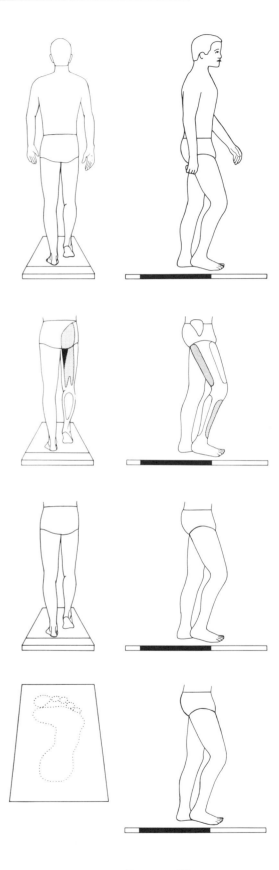

Frame = 23
Time = 0.88 s

Company Name:	Novel Electronics, Inc.	
Address:	3817 Garfield Avenue South	Beichstrasse 8 8000 Munchen 40
	Minneapolis, MN 55409 USA	West Germany
Telephone:	(612) 822-9912	(089) 390102
Product Name:	EMED Systems	
Price Range:	C (Mini EMED) E (EMED SF and Micro EMED)	

The EMED systems—the pressure mat and pressure insole—are based on the same capacitive technology and may be driven by the same integrated electronics and computer plus software. The insole has 85 separate transducer sites (each about 18 mm × 18 mm); the pressure mat has an area of 0.36 m × 0.18 m and up to 2,048 transducer sites (each about 6 mm × 6 mm). The sampling rate is 16 frames/s for the Mini EMED and up to 100 frames/s for the other two systems; the pressures are color-coded in seven different color values, similar to thermographic analysis, and have a numerical value between 0.02 and 1.27 MPa. Unfortunately, color cannot be printed in this manual because of the prohibitive cost, but this technique clearly enhances understanding and pinpoints areas of unduly high or low pressure. The advantages of these systems are the intuitive method of displaying pressure data, good spatial resolution and pressure range, simplicity of operation and understanding, and the capability to switch between a pressure mat and an insole for the EMED SF and Micro EMED systems. The disadvantages are the poor sampling rate for the mini EMED, the small size of platform, which could lead to targeting, encumbrance of the subject by the trailing wires of the pressure insole system, and the lack of shear force information (i.e., no anterior/posterior or medial/lateral forces).

Company Name:	Infotronic Medical Engineering
Address:	PO Box 73
	7650 AA Tubbergen
	Netherlands
Telephone:	(053) 893478
Product Name:	Computer Dyno Graph (CDG)
Price Range:	C

The CDG consists of the following components: a pair of pressure insoles, each having eight capacitive transducers (30 mm × 30 mm × 1.5 mm); a battery-powered measuring unit which is worn either around the subject's neck or waist and records 16 channels at 50 Hz; an interface cable based on the parallel centronics standard; a computer system based on the Atari personal computer; and an analysis program complete with pull-down menus and windows. A total of 20 s can be stored in the measuring unit's memory before connecting the unit to the host computer for downloading the data. This frees the subject from

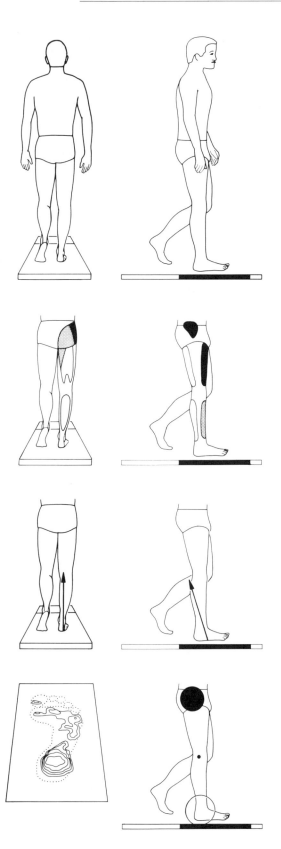

Frame = 6
Time = 0.20 s

being tied to the base station by trailing wires. The CDG transducers have a range of 0.004 to 1.11 MPa, and the measuring unit has a resolution of 0.004 MPa (i.e., 8 bits, or 256 levels). The advantages of the system are the versatility of the software, the relative lack of encumbrance of the subject, and good temporal resolution and pressure range. Its disadvantages are poor spatial resolution and placement of the pressure transducers on the insole, a microcomputer which is not really a standard in scientific laboratories, and the lack of shear force information.

Company Name:	Human Performance Technologies, Inc.
Address:	12 Technology Drive, Suite 6
	East Setauket, NY 11733
	USA
Telephone:	(516) 689-6521
Product Name:	2P-QDA
Price Range:	B

This system consists of six pairs of foot pressure insoles (in assorted sizes) which indicate 14 zones of pressure, 7 per insole. Also included with the system is an IBM PC/XT plus appropriate peripherals. The pressure transducers work on the capacitive principle, and though the company claims that they can be calibrated, no specifications of operating range or frequency response are provided in the product literature. The software allows an oblique view of the relative pressure magnitudes at the 14 discrete locations; force-time curves for each location, color-coded for identification purposes, but with no force or pressure scale; temporal information on foot contact; and simultaneous capture of electromyographic data. The system is suitable for detecting left/right asymmetries during gait and for studying the effects of shoe orthotics. Its advantages are its portability and ability to gather EMG data. Because it fits inside the shoe, it avoids the targeting problem. However, there are a number of major disadvantages: Only 7 transducers, in fixed positions, cover the whole foot; only 6 different insole sizes are provided, thus precluding an adequate match for the individual foot; the insoles are relatively thick (about 5 mm); load range and frequency response are unknown; and the subject is encumbered by trailing wires.

Company Name:	Tekscan
Address:	451 D Street
	Boston, MA 02110
	USA
Telephone:	(617) 737-8734
Product Name:	F-Scan
Price Range:	C

The F-Scan system uses a fairly new sensor technology, called force sensing resistors, that enables up to 1,000 separate sensing sites to be monitored in a thin (0.1 mm) transducer placed inside the shoe. The transducers, which are disposable, can be cut to fit the size of the patient's shoe; though this reduces the number of sensing sites, the system has a constant distribution of approximately 3 sensors/cm^2. The transducers are attached to the computer via an umbilical cord of cables, but a wireless unit will soon be released. Included with the base system are the driving software, a high-resolution color monitor for displaying a 3-D movie of foot pressure in real-time, a color printer for hard-copy printouts, and 20 transducers. The major advantages of the system are that the transducer can be customized to fit inside any shoe, which avoids the targeting problem, data are available in real-time, and the use of color for pressure levels enhances understanding. However, there are some disadvantages: The loading response is nonlinear, and because each sensing site is not separately calibrated, the accuracy of the system is unknown. Also, the durability of the flexible transducers is unknown, and the subject is encumbered by trailing wires.

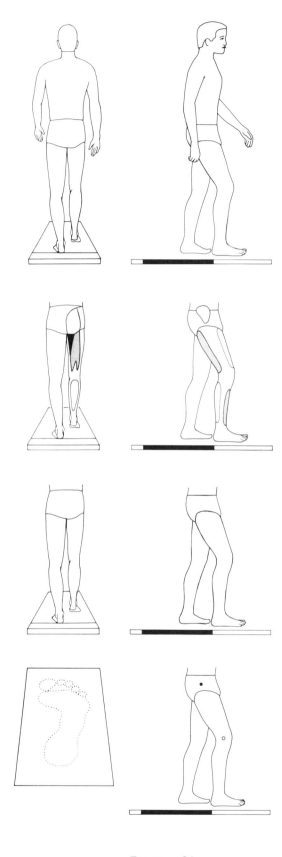

Frame = 24
Time = 0.92 s

Company Name: WM Automation
Address: 76 Bally Duff Rd., Carnmoney
                Newton Abbey, County Antrim
                Northern Ireland BT36 6PB
Telephone: (352) 72011
Product Name: Musgrave Footprint
Price Range: E

The Musgrave Footprint system utilizes force-sensing resistors (FSRs) to measure the pressure between the bottom of a subject's foot and the ground. Two separate pressure plates, each about 40 cm × 20 cm in size and containing 2,048 individual sensors, capture the vertical ground reaction forces from the left and right feet. The sensors can measure up to 1.5 MPa, and the temporal resolution is about 60 Hz. The software makes extensive use of color to illustrate the pressure profiles on each foot, to highlight the pressure on designated areas of the foot, to track the center of pressure, and to provide a summation of the total vertical load, which is expressed in kilograms. The advantages of this system are the simplicity of operation, the intuitive method of displaying pressure data, good spatial and pressure resolution, and portability. The disadvantages are the relatively small platform size, which could lead to targeting, the lack of shear information, limited temporal resolution, and unknown accuracy across separate transducer sites.

Integrated Software Packages

In many ways, a good integrated software package is the glue that holds all the disparate parts of gait analysis together. It is not sufficient merely to collect data from various instruments and then examine the results in isolation. The aim of gait analysis is to combine the data from electromyography, anthropometry, displacement of segments, and ground reaction forces in a meaningful and mechanically sound manner. This is what we have done in *GaitLab*. You can do the same: *GaitLab* includes a copy of our software, as well as directions on how to run it. In

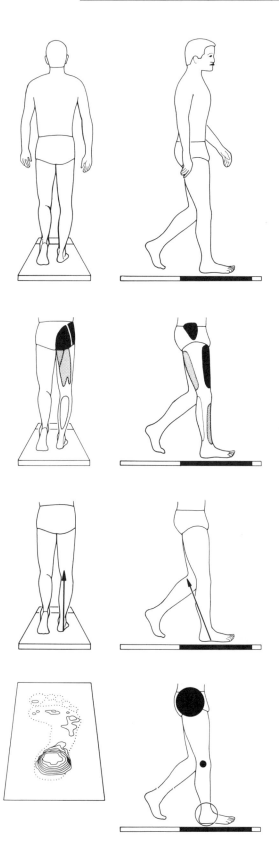

Frame = 5
Time = 0.16 s
Left toe-off

this section, we describe three products currently in use at a number of clinical gait laboratories.

Company Name: Motion Analysis Corporation
Address: 3650 North Laughlin Road
Santa Rosa, CA 95403
USA
Telephone: (707) 579-6511 or 1-800-862-1338
Product Name: OrthoTrak
Price Range: C

OrthoTrak was developed in conjunction with professionals at the Cleveland Clinic Foundation in Cleveland, OH, and the University of Florida in Gainesville, FL. It is especially designed to be used with the 3-D ExpertVision system from Motion Analysis, which was described earlier in this appendix. Some of OrthoTrak's key features are as follows: a 25-marker system that defines the position of all body segments and particularly the lower extremities in great detail; stick figure animation (either forward or backward) that may be performed with any one of three views (side, front, above); calculation of internal joint centers from external marker positions; integrated 32-channel analog data subsystem for managing EMG, force plate, and other signals; intuitive user interface with mouse and pull-down menus; and a facility with which the user can design his or her own report format. The advantages are its versatility; its ease of use; the integration of kinematic, force plate, and EMG data; and the emphasis on joint forces and moments (i.e., the cause of the observed motion). Its disadvantages are the large number of markers, 25, which challenges both the tracking algorithm and the operator's patience, and the use of a bulky triad of markers in the midthigh and midcalf regions. Recent versions of the software have accommodated a simpler set of 15 markers (Kadaba et al., 1990).

Company Name: ADTECH
Address: 2002 Ruatan St.
Adelphi, MD 20783
USA
Telephone: (301) 434-0785
Product Name: AMASS and ADG
Price Range: C (AMASS)
B (ADG)

The ADTECH Motion Analysis Software System (AMASS) is designed to be used with a video-based motion analysis system such as VICON and has the following features: corrections for nonlinearities due to lens, detector, and electronics of the cameras, yielding an accuracy of 1 to 3 mm within a 2 m^3 volume; automatic identification of 3-D calibration reference markers, which allows a calibration to be performed in less than 5 minutes; automatic tracking of 3-D position of markers (rather than the traditional 2-D approach for each camera), which decreases data reduction time by almost an order of magnitude; and flexible camera placement. The 3-D tracking algorithm and interactive trajectory identification routine are its biggest advantages and have enabled laboratories equipped with the VICON system to reduce their turn-around time considerably.

The ADTECH Graphics (ADG) program is a general purpose software package, designed for the convenient manipulation and pictorial presentation of 3-D kinematic and digitized analog data. Developed 5 years ago for use with the Selspot system, it is now available for use with the VICON system from Oxford Metrics. Its major advantage is its versatility and ease of use. Although the ADG program can compute many kinematic variables and display them with force plate and EMG data in a single plot, it does not possess the capability of calculating kinetic data such as joint forces and moments.

Company Name: OsteoKinetics Corporation
Address: 18 Crosby Road
Newton, MA 02167
USA
Telephone: (617) 332-5954
Product Name: BioTRACK and GraphTRACK
Price Range: B (BioTRACK)
A (GraphTRACK)

These two software packages require an IBM PC/AT or compatible computer and a graphics card (CGA, EGA, VGA, or Hercules), and support a variety of hard-copy devices. BioTRACK is the kinematic engine that processes the raw displacement data, whereas GraphTRACK is used to plot the processed data in a variety of formats. The movement of body segments is measured by the use of light-emitting diodes (LEDs) attached to a lightweight frame that is strapped to the segment of interest. BioTRACK provides information on the 3-D location of each LED and its time derivatives (velocity and acceleration). The user can specify interpolation and filtering options for further processing of the raw data. The program also calculates and stores the six degrees of freedom of each link segment, as well as calculating the joint rotations (i.e., the position of one segment relative to another). GraphTRACK, using self-explanatory menus, allows the user to specify up to six different graphs on a single screen and multiple curves in each graph. Further analyses can be done in the time and frequency domains, the latter option providing the user with the necessary insight for selecting optimal filtering schemes for data smoothing. In addition, GraphTRACK enables the user to plot one variable against any other variable (e.g., hip angle vs. knee angle, or position in terms of velocity), thus providing further insight into function and control mechanisms in human movement. The major disadvantage of these two software packages is that they deal only with kinematic data and do not integrate other important movement parameters such as anthropometry, electromyography, or ground reaction forces.

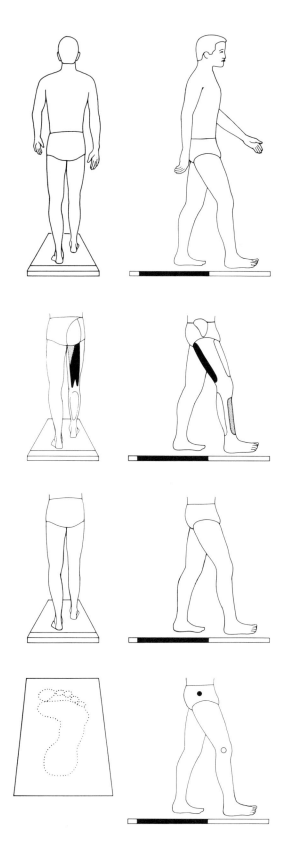

Frame = 25
Time = 0.96 s

126

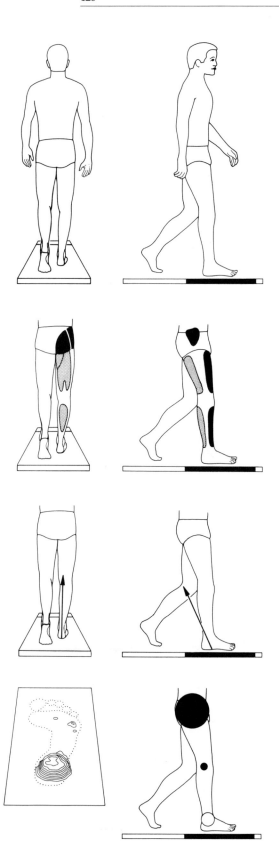

Frame = 4
Time = 0.12 s

References

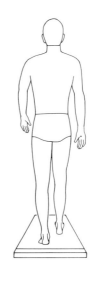
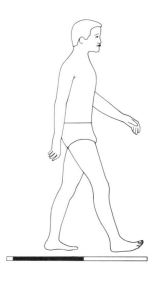

Ackland, T.R., Blanksby, B.A., & Bloomfield, J. (1988). Inertial characteristics of adolescent male body segments. *Journal of Biomechanics*, **21**, 319-327.

Andriacchi, T.P., & Strickland, A.B. (1985). Gait analysis as a tool to assess joint kinetics. In N. Berme, A.E. Engin, & K.M. Correia da Silva (Eds.), *Biomechanics of normal and pathological human articulating joints* (pp. 83-102). The Hague: Martinus Nijhoff.

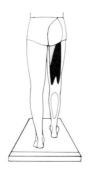

Apkarian, J., Naumann, S., & Cairns, B. (1989). A three-dimensional kinematic and dynamic model of the lower limb. *Journal of Biomechanics*, **22**, 143-155.

Atha, J. (1984). Current techniques for measuring motion. *Applied Ergonomics*, **15**, 245-257.

Basmajian, J.V., & DeLuca, C.J. (1985). *Muscles alive: Their functions revealed by electromyography*. Baltimore: Williams & Wilkins.

Bechtol, C.O. (1975). Normal human gait. In J.H. Bowker & C.B. Hall (Eds.), *Atlas of orthotics: American Academy of Orthopaedic Surgeons* (pp. 133-143). St. Louis: Mosby.

Bernstein, N. (1967). *The coordination and regulation of movements*. London: Pergamon Press.

Brand, R.A., & Crowninshield, R.D. (1981). Comment on criteria for patient evaluation tools. *Journal of Biomechanics*, **14**, 655.

Braune, W., & Fischer, O. (1889). Uber den Schwerpunkt des menschlichen Korpers [On the center of gravity of the human body]. *Abhandlungen der Mathematisch—Physische Klasse der Konigl. Sachsischen Gesellschaft der Wissenschaften*, **15**, 561-572. (Leipzig: Hirzel)

Brooks, C.B., & Jacobs, A.M. (1975). The gamma mass scanning technique for inertial anthropometric measurement. *Medicine and Science in Sports*, **7**, 290-294.

Campbell, K.R., Grabiner, M.D., Hawthorne, D.L., & Hawkins, D.A. (1988). The validity of hip joint center predictions from anatomical landmarks. *Journal of Biomechanics*, **21**, 860.

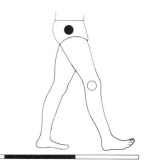

Frame = 26
Time = 1.00 s
Right deceleration

REFERENCES

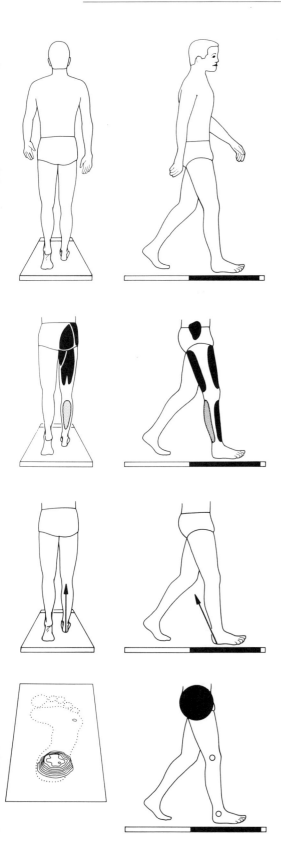

Frame = 3
Time = 0.08 s
Right foot-flat

Cappozzo, A. (1984). Gait analysis methodology. *Human Movement Science*, **3**, 27-50.

Cavanagh, P.R. (1987). On "muscle action" versus "muscle contraction." *Journal of Biomechanics*, **21**, 69.

Cavanagh, P.R. (1988). Studies in the biomechanics of distance running and plantar pressure distribution. In G. de Groot, A.P. Hollander, P.A. Huijing, & G.J. van Ingen Schenau (Eds.), *Biomechanics XI-A* (pp. 3-14). Amsterdam: Free University Press.

Cavanagh, P.R., Hennig, E.M., Bunch, R.P., & Macmillan, N.H. (1983). A new device for measurement of pressure distribution inside the shoe. In H. Matsui & K. Kobayashi (Eds.), *Biomechanics VIII-B* (pp. 1089-1096). Champaign, IL: Human Kinetics.

Chandler, R.F., Clauser, C.E., McConville, J.T., Reynolds, H.M., & Young, J.W. (1975). *Investigation of inertial properties of the human body* (Aerospace Medical Research Laboratory Tech. Rep. No. 74-137). Dayton, OH: Wright-Patterson Air Force Base, AMRL. (Prepared for U.S. Department of Transportation, National Highway Traffic Safety Administration, Contract No. DOT-HS-017-2-315-1A; National Technical Information Service No. AD-A016 485)

Chao, E.Y. (1980). Justification of triaxial goniometer for the measurement of joint rotation. *Journal of Biomechanics*, **13**, 989-1006.

Cochran, G.V.B. (1982). *A primer of orthopaedic biomechanics*. New York: Churchill Livingstone.

Crowninshield, R.D., & Brand, R.A. (1981). A physiologically based criterion of muscle force prediction in locomotion. *Journal of Biomechanics*, **14**, 793-801.

Dainis, A. (1980). Whole body and segment center of mass determination from kinematic data. *Journal of Biomechanics*, **13**, 200.

Davy, D.T., & Audu, M.L. (1987). A dynamic optimization technique for predicting muscle forces in the swing phase of gait. *Journal of Biomechanics*, **20**, 187-202.

Dempster, W.T. (1955). *Space requirements of the seated operator: Geometrical, kinematic, and mechanical aspects of the body with special reference to the limbs* (Wright Air Development Center Tech. Rep. No. 55-159). Dayton, OH: Wright-Patterson Air Force Base, WADC. (National Technical Information Service No. AD-087892)

Enoka, R.M. (1988). *Neuromechanical basis of kinesiology*. Champaign, IL: Human Kinetics.

Erdmann, W.S. (1989). Application of computerized tomography for obtaining inertia quantities of the human trunk. *Journal of Biomechanics*, **22**, 1100.

Fenn, W.O. (1930). Work against gravity and work due to velocity changes in running. *American Journal of Physiology*, **93**, 433-462.

Fried, I. (1973). *The chemistry of electrode processes*. New York: Academic Press.

Gainey, J.C., Kadaba, M.P., Wootten, M.E., Ramakrishnan, H.K., Siris, E.S., Lindsay, R., Canfield, R., & Cochran,

REFERENCES

G.V.B. (1989). Gait analysis of patients who have Paget disease. *Journal of Bone and Joint Surgery*, **71-A**, 568-579.

Geddes, L.A. (1972). *Electrodes and the measurement of bioelectric events*. New York: Wiley-Interscience.

Goldstein, H. (1965). *Classical mechanics*. Reading, MA: Addison-Wesley.

Gould, P. (1985). *The geographer at work*. London: Routledge & Kegan Paul.

Greaves, J.O.B. (1990). Recent software developments for biomechanical assessment. In J.S. Walton (Ed.), *SPIE Symposium: Vol. 1356. Mini-symposium on image-based motion measurement* (pp. 8-11). Bellingham, WA: Society of Photo-Optical Instrumentation Engineering.

Greenwood, D.T. (1965). *Principles of dynamics*. Englewood Cliffs, NJ: Prentice Hall.

Grieve, D.W. (1968). Gait patterns and the speed of walking. *Biomedical Engineering*, **3**, 119-122.

Grood, E.S., & Suntay, W.J. (1983). A joint coordinate system for the clinical description of three-dimensional motions: Application to the knee. *Journal of Biomechanical Engineering*, **105**, 136-144.

Hanavan, E.P., Jr. (1964). *A mathematical model of the human body* (Aerospace Medical Research Laboratories Tech. Rep. No. 64-102). Dayton, OH: Wright-Patterson Air Force Base, AMRL. (NTIS No. AD 608463)

Hatze, H. (1980). A mathematical model for the computational determination of parameter values of anthropomorphic segments. *Journal of Biomechanics*, **13**, 833-843.

Hinrichs, R. (1985). Regression equations to predict segmental moments of inertia from anthropometric measurements. *Journal of Biomechanics*, **18**, 621-624.

Hof, A.L. (1988). Assessment of muscle force in complex movements by EMG. In G. de Groot, A.P. Hollander, P.A. Huijing, & G.J. van Ingen Schenau (Eds.), *Biomechanics XI-A* (pp. 111-117). Amsterdam: Free University Press.

Huang, H.K., & Wu, S.C. (1976). The evaluation of mass densities of the human body in vivo. *Computers in Biology and Medicine*, **6**, 337-343.

Inman, V.T., Ralston, H.J., & Todd, F. (1981). *Human walking*. Baltimore: Williams & Wilkins.

Jacobsen, B., & Webster, J.G. (1977). *Medicine and clinical engineering*. Englewood Cliffs, NJ: Prentice Hall.

Jensen, R.K. (1986). Body segment mass, radius, and radius of gyration properties in children. *Journal of Biomechanics*, **19**, 355-368.

Kadaba, M.P., Ramakrishnan, H.K., & Wootten, M.E. (1990). Measurement of lower extremity kinematics during level walking. *Journal of Orthopaedic Research*, **8**, 383-392.

Kaiser, H.F. (1961). A note on Guttman's lower bound for the number of common factors. *British Journal of Mathematical and Statistical Psychology*, **14**, 1-2.

Keele, K.D. (1983). *Leonardo da Vinci's Elements of the Science of Man*. New York: Academic Press.

Kim, J.O., & Kohout, F.J. (1975). Multiple regression analysis:

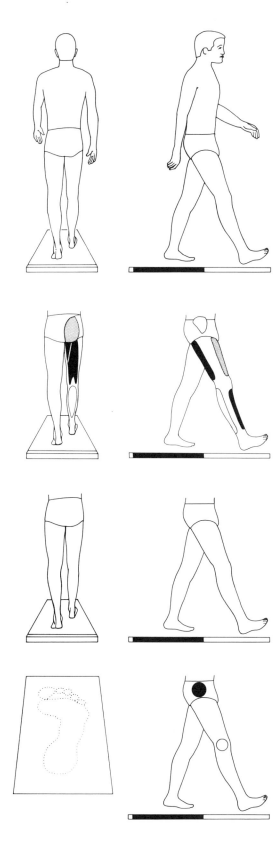

Frame = 27
Time = 1.04 s

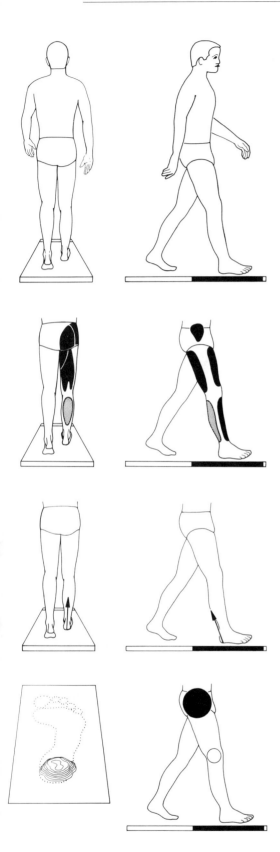

Frame = 2
Time = 0.04 s

Subprogram regression. In N.H. Nie (Ed.), *Statistical package for the social sciences*. New York: McGraw-Hill.

Lambrinudi, C. (1927). New operation on drop-foot. *British Journal of Surgery*, **15**, 193.

Lanshammar, H. (1985). Measurement and analysis of displacement. In H. Lanshammar (Ed.), *Gait analysis in theory and practice* (pp. 29-45). Uppsala, Sweden: Uppsala Universitet, UPTEC 85R.

Loeb, G.E., & Gans, C. (1986). *Electromyography for experimentalists*. Chicago: University of Chicago Press.

Loeb, G.E., & Levine, W.S. (1990). Linking musculoskeletal mechanics to sensorimotor neurophysiology. In J.M. Winters & S.L.Y. Woo (Eds.), *Multiple muscle systems* (pp. 165-181). New York: Springer-Verlag.

Mann, R.A., & Hagy, J. (1980). Biomechanics of walking, running, and sprinting. *American Journal of Sports Medicine*, **8**, 345-350.

Marey, E.J. (1886). *La machine animal*. Felix Alcan (Ed.), Paris.

Martin, P.E., Mungiole, M., Marzke, M.W., & Longhill, J.M. (1989). The use of magnetic resonance imaging for measuring segment inertial properties. *Journal of Biomechanics*, **22**, 367-376.

Miller, D.I., & Nelson, R.C. (1973). *Biomechanics of sport*. Philadelphia: Lea & Febiger.

Murray, M.P., & Gore, D.R. (1981). Gait of patients with hip pain or loss of hip joint motion. In J. Black & J.H. Dumbleton (Eds.), *Clinical biomechanics: A case history approach*. New York: Churchill Livingstone.

Patla, A.E. (1985). Some characteristics of EMG patterns during locomotion: Implications for the locomotor control process. *Journal of Motor Behavior*, **17**, 443-461.

Pezzack, J.C., Norman, R.W., & Winter, D.A. (1977). An assessment of derivative determining techniques used for motion analysis. *Journal of Biomechanics*, **10**, 377-382.

Procter, P., & Paul, J. (1982). Ankle joint biomechanics. *Journal of Biomechanics*, **15**, 627-634.

Radar, C.M., & Gold, B. (1967). Digital filtering design techniques in the frequency domain. *Proceedings of IEEE* (Institute of Electrical and Electronic Engineers), **55**, 149-171.

Rieger, P.H. (1987). *Electrochemistry*. Englewood Cliffs, NJ: Prentice Hall.

Solomon, C. (1989). *Enchanted drawings: The history of animation*. New York: Knopf.

Sutherland, D.H. (1984). *Gait disorders in childhood and adolescence*. Baltimore: Williams & Wilkins.

Sutherland, D.H., Olshen, R.A., Biden, E.N., & Wyatt, M.P. (1988). *The development of mature walking*. Oxford, England: Mac Keith Press.

Synge, J.L., & Griffith, B.A. (1959). *Principles of mechanics*. New York: McGraw-Hill.

Tylkowski, C.M., Simon, S.R., & Mansour, J.M. (1982). Internal rotation gait in spastic cerebral palsy. *The Hip: Proceedings of 10th Open Meeting of the Hip Society*, **10**, 89-125.

van den Bogert, A.J. (1989). *Computer simulation of locomotion in the horse*. Doctoral dissertation (ISBN 90-9003176-6), University of Utrecht, Netherlands.

REFERENCES

Vaughan, C.L. (1982). Smoothing and differentiation of displacement-time data: An application of splines and digital filtering. *International Journal of Bio-Medical Computing*, **13**, 375-386.

Vaughan, C.L. (1983). Forces and moments at the hip, knee, and ankle joints. *Annual Report of the Oxford Orthopaedic Engineering Centre*, **10**, 17-18.

Vaughan, C.L., Andrews, J.G., & Hay, J.G. (1982). Selection of body segment parameters by optimization methods. *Journal of Biomechanical Engineering*, **104**, 38-44.

Vaughan, C.L., du Toit, L.L., & Roffey, M. (1987). Speed of walking and forces acting on the feet. In B. Jonsson (Ed.), *Biomechanics X-A* (pp. 349-354). Champaign, IL: Human Kinetics.

Vaughan, C.L., Hay, J.G., & Andrews, J.G. (1982). Closed loop problems in biomechanics. Part I—A classification system. *Journal of Biomechanics*, **15**, 197-200.

Walton, J.S. (1990). *SPIE Symposium: Vol. 1356. Mini-symposium on image-based motion measurement*. Bellingham, WA: Society of Photo-Optical Instrumentation.

Warner, A. (1972). *Modern biopotential electrode principles and applications*. Fullerton, CA: Beckman Instruments.

Webster, J.G. (1976). *Medical instrumentation: Application and design*. Madison: University of Wisconsin.

Webster's Ninth New Collegiate Dictionary. (1983). Springfield, MA: Merriam-Webster.

Winter, D.A. (1979). *Biomechanics of human movement*. New York: Wiley.

Winter, D.A. (1987). *The biomechanics and motor control of human gait*. Waterloo, ON, Canada: University of Waterloo.

Winter, D.A., Rau, G., Kadefors, R., Broman, H., & DeLuca, C.J. (1980). Units, terms, and standards in the reporting of EMG research. Report by the International Society of Electrophysiological Kinesiology.

Woltring, H.J. (1984). On methodology in the study of human movement. In H.T.A. Whiting (Ed.), *Human motor actions—Bernstein reassessed* (pp. 35-73). Amsterdam: Elsevier.

Wood, G.A., & Jennings, L.S. (1979). On the use of spline functions for data smoothing. *Journal of Biomechanics*, **12**, 477-479.

Wootten, M.E., Kadaba, M.P., & Cochran, G.V.B. (1990a). Dynamic electromyography: I. Numerical representation using principal component analysis. *Journal of Orthopaedic Research*, **8**, 247-258.

Wootten, M.E., Kadaba, M.P., & Cochran, G.V.B. (1990b). Dynamic electromyography: II. Normal patterns during gait. *Journal of Orthopaedic Research*, **8**, 259-265.

Yeadon, M.R., & Morlock, M. (1989). The appropriate use of regression equations for the estimation of segmental inertia parameters. *Journal of Biomechanics*, **22**, 683-689.

Zatsiorsky, V.M., & Seluyanov, V. (1985). Estimation of the mass and inertia characteristics of the human body by means of the best predictive regression equations. In D.A. Winter, R.W. Norman, R.P. Wells, K.C. Hayes, & A.E. Patla (Eds.), *Biomechanics IX-B* (pp. 233-239). Champaign, IL: Human Kinetics.

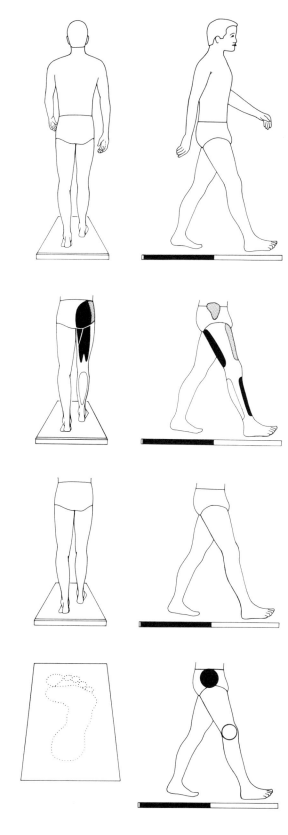

Frame = 28
Time = 1.08 s

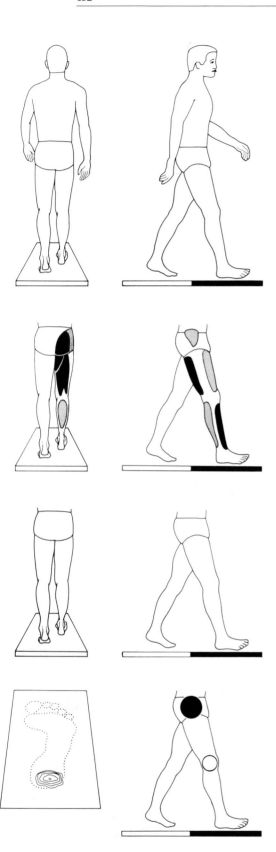

Frame = 1
Time = 0.00 s
Right heel strike

Index

A
Abduction, 20, 32, 33, 41, 93-95
Acceleration
 angular, 32, 33-35, 36, 55, 96
 and calculating centers of gravity, 29-31, 93
 during gait cycle, 11, 54
Action and reaction, law of, 40
Addison Gallery (Andover, MA), 78
Adduction, 20, 32, 33, 41, 93-95
Adductor longus, 59, 69, 71
Adductor magnus, 53
ADG (ADTECH Graphics), 124-125
ADTECH, 115, 124-125
Advanced Mechanical Technology, Inc. (AMTI), 65, 119
Aliasing, 49
AMASS (ADTECH Motion Analysis Software System), 124-125
Amplification, of electrode signals, 51
Anatomical position, standard, 7
Andrews, J. G., 4
Angle Display Unit (Penny & Giles Blackwood, Ltd.), 111-112
Animals in Motion (Muybridge), 78
Animation sequences, dynamic, 77-81
Ankle. *See also* Foot
 of clinical subject, 73-74
 control of, 1
 joint forces and moments at, 38-40
 plantar flexion and dorsiflexion in, 55
 predicting position of joint center, 25-26, 88
 reference frame for, 32, 89-90
Ankle equinus, 11
Anterior superior iliac spines (ASIS)
 breadth of, 17-18
 and predicting hip joint center, 26-27
Anthropometric Measurements Set (Carolina Biological Supply Company), 109
Anthropometry, 4, 15, 17-18, 19
 of clinical subject, 64, 65-66
 equipment for, 109
Arthrodesis, 63
Atha, J., 109
Avascular necrosis, 10
Axial tomography, 16-17

B
Balance, 12, 67
Baltimore Therapeutic Equipment Co., 119-120
Bioengineering Technology & Systems, 115
BioTRACK (OsteoKinetics Corporation), 125

BIOVEC-1000 Force Platform System, 119
Bipedalism, 8, 79
Biphasic factor, 59-60
Body, human
 center of gravity of, 10
 mass of, 16, 17, 18-20
 planes of, 7-8
  coronal (or frontal), 7, 80-81
  sagittal, 7, 25, 67, 79-80
  transverse, 7
Body segment parameters
 and anthropometry, 17-18, 86-87
 definition of, 16
 Gaitmath mathematics for, 83-87
 prediction of segment mass, 16, 17, 18-20
 prediction of segment moments of interia, 16, 20-22
 problems in estimation of, 16-17
 selection of segments, 16
Bunch, R. P., 78
Butterworth filter, second-order low-pass, 91

C
Cadaver averages, 16, 17
Calf. *See also* Knee
 center of gravity of, 32-33, 34, 91
 circumference of, 17-18
 Euler angles of, 33, 35
 joint dynamics of, 100-101, 103-104
 length of, 17-18
 markers for kinematic measurement of, 26, 88-89
 orientation of, 28-29
 predicting mass of, 18-20
 predicting position of knee joint center, 26
Calf circumference, 17-18
Calf length, 17-18
Calipers
 beam, 17, 18, 64
 sliding, 18
 special, 109
 spreading, 18
Carolina Biological Supply Company, 109
Cavanagh, P.R., 49, 78
Center for Locomotion Studies (CELOS), 79
Centers of gravity. *See* Gravity, centers of
Cerebellar ataxia, 12
Cerebral palsy
 impact on central nervous system, 3
 measuring ground reaction force of patient with, 13
 types of

Cerebral palsy (*continued*)
  athetoid form of, 12, 63
  with hypotonia, 63
  spastic, 11
Chandler, R.F., 19, 83-87
Chao, E.Y., 32, 93
Charcot-Marie-Tooth disease, 3, 63, 74
Chattecx Corporation, 110-111
Clauser, C.E., 19, 83-87
Cleveland Clinic Foundation, 124
Clinical analysis, case study for, 64-76
Cochran, G.V.B., 60
Columbus Instruments, 116
Computer Dyno Graph (Infotronic Medical Engineering), 121-122
Coronal (or frontal) plane, 7, 80-81

D
Data Recorder (Penny & Giles Blackwood, Ltd.), 111-112
Deceleration, 11, 54, 55
Digital Equipment PDP/VAX graphics terminal, 114, 115
Digital filter algorithm, 30, 31, 91, 96
Displacement
  as component in walking, 3
  equipment for measuring, 109-118
  measurement of, 4, 12-13
Distance, measurement of, 11-12
Dorsiflexion, 32, 55, 63, 73, 95
Double layer, 46-47
du Toit, L.L., 72
Dynamic Pedobarograph (Baltimore Therapeutic Equipment Co.), 119-120

E
Electrode potential, 46
Electrodes
  advantages and disadvantages of, 50-51
  definition of, 45-48
  skin surface preparation for, 50, 65
  types of, 50-51
Electromyography (EMG), 4, 45-61
  of clinical subject, 65, 67-71
  electrochemical principles, 45-48
  electrophysiology, 48-49
  equipment for, 108-109
  measuring muscle activity by, 13-14
  and muscle interactions, 55-61
  and phasic behavior of muscles, 52-55
  sampling frequency in, 49-50
  signal processing methods for, 51-52
  surface versus indwelling electrodes for, 50-51, 61
Electrophysiology, 48-49
Elite Motion Analyser (Bioengineering Technology & Systems), 115
EMED Systems (Novel Electronics, Inc.), 121
Equipment for gait analysis
  for anthropometry, 109
  cost of, 107
  for electromyography, 108-109
  integrated software packages for, 123-125
  for measuring displacement of segments, 109-118
  for measuring ground reaction forces, 118-123
Erector spinae, 59, 67-68
Etherbox (Oxford Metrics Ltd.), 114
Ethernet Local Area Network, 114
Euler angles, 22, 32-33, 95-96
Eversion, 95
ExpertVision (Motion Analysis Corporation), 113-114, 124
Extension, 20, 32, 33, 41, 93-95
Extensor digitorum longus, 52, 55, 58, 59
Extensor hallucis longus, 55
External forces, as component in walking, 2-3

F
Factor analysis, 58-59
Feedback, sensory, 2
Femoral epicondyle, 26
Fenn, W.O., 118
First double support phase, 9-10
Flexion, 20, 32, 33, 41, 93-95
Foot. *See also* Ankle; Footprints; Heel
  breadth of, 17-18
  center of gravity of, 30-31, 91
  and events in gait cycle, 10-11
  Free Body Diagram (FBD) for, 38, 39
  joint dynamics of, 99-100, 102-103
  length of, 17-18
  markers for kinematic measurement of, 25-26, 88
  orientation of, 29
  position during walking, 8, 9-11
  predicting mass of, 18-20
  predicting position of ankle joint center, 25-26
  pressure distribution under, 80
  and rotation of body, 12
  slapping of, 71, 73
  and torso movement, 80
Foot breadth, 17-18
Foot-flat, 10-11, 54
Foot length, 17-18
Footprints, and distance measurement, 11-12
FootTrak (Motion Analysis Corporation), 113-114
Force plates, 36, 40, 65, 72, 118-120
Free Body Diagram (FBD), 38-39
Frontal plane. *See* Coronal (or frontal) plane
F-Scan (Tekscan), 122-123

G
Gait (human)
  clinical example of, 63-76
  computer simulation of, 78-79
  cyclic nature of, 8-12
  dynamic animation sequences for, 77-132 (margins)
  equipment for analysis of, 107-125
  integration of components in, 5
  measurement of, 4-6, 15-43
  parameters of, 12-14
  pathological, 3-4, 63-76
  periodicity of, 8-12
  sequence of events in, 3
  symmetry in, 10
  as three-dimensional, 7-8
  top-down analysis of, 2-4
Gait analysis, science of, 77-81
Gait Analysis System (MIE Medical Research Limited), 110
Gait cycle
  asymmetry in, 10, 11, 12
  distance measures for, 11-12
  and EMG patterns for muscles of lower extremities, 52, 53
  events of, 10-11
  of patients with pathologies, 11
  phases of, 9-10, 11, 22, 80
GaitLab, 123-124
  animated muscle sequences in, 53
  Animate program of, 81
  animation figures in, 79
  DIGFIL subroutine in, 91
  digital filter algorithm in, 30, 31
  FIDIFF subroutine in, 93
  integration of measurements in, 5-6
  KIN files in, 5, 24, 65
  linear envelope method of representing EMG signals in, 52, 65
  Man.ANG file in, 6, 32, 34
  Man.DYN file in, 6, 41, 42
  Man.FPL file in, 5, 36, 37, 40, 65
  Man.JNT file in, 5, 27, 28
  Man.REF file in, 5, 91
  muscle interactions explored in, 56
  structure of data files in, 5
Gaitmath, 5, 6, 17, 21, 83-106
Gaitplot, 5, 6
Gamma ray scanning, 16-17
Gans, C., 50, 51, 108

Gastrocnemius, 57, 61
Geddes, L. A., 50
Gluteus maximus, 52, 57, 59, 67, 74
Gluteus medius, 52, 55, 56, 57, 59, 68-69
Gold, B., 91
Goldstein, H., 33, 96
Goniometer System (Penny & Giles Blackwood, Ltd.), 111-112
Gould, P., 57
GP-8-3D Digitizer (Science Accessories Corporation), 112
GraphTRACK (OsteoKinetics Corporation), 125
Gravity, centers of (CG), 15, 16, 29-31
    Gaitmath mathematics for, 91-93
Greaves, J.O.B., 79
Griffith, B.A., 33, 96
Grood, E.S., 32, 93
Ground reaction forces
    as component in walking, 2, 3, 8
    equipment for measuring, 118-123
    measurement of, 4, 13, 36-38
    orientation of, 80
    relationship to joint moments of inertia, 80

H
Hagy, J., 55
Half-cell potentials, 46-47, 51
Hamstrings, 53, 56, 58, 59, 68, 69, 70, 74
Hay, J.G., 4
Heel, 25
Heel-off, 10-11, 54
Heel strike, 10-11, 54, 55, 58, 60
Heel strike factor, 59, 60
Hennig, E.M., 78
High Speed Photographic & Videographic Motion Analysis Systems (Instrumentation Marketing Corporation), 112-113
Hinrichs, R., 87
Hip. *See also* Pelvis
    anatomical joint angles of, 32, 34
    of clinical subject, 74, 76
    joint forces and moments at, 38, 40
    pain in, 7, 10
    predicting position of joint center, 26-27, 89
    reference frame for, 32, 89-90
Hip flexor muscles, 11, 13
Hip pain
    bilateral, 7
    unilateral, 10
Hof, A.L., 45
Horner, William, 78
Human Figure in Motion, The (Muybridge), 78
Human Performance Technologies, Inc., 115-116, 122

I
IBM PC, PC/AT, PC/XT, and PS/2 personal computers, 108, 110-120, 122, 125
Inertia. *See* Moments of inertia
Infotronic Medical Engineering, 121-122
Initial contact, 11, 54
Initial swing, 11, 54
Inman, V.T., 1, 52, 53, 55, 78
Instrumentation Marketing Corporation, 112-113
International Society of Biomechanics, 78
International Society of Electrophysiological Kinesiology (ISEK), 49, 51
Inversion, 95

J
Joint angles, anatomical, 32, 93-95
Joint force, relationship to moments of inertia, 40-43
Joints, dynamics of, 15, 35-43
    calculation of joint forces and moments, 38-40
    expression of joint forces and moments, 40-43
    Gaitmath mathematics for, 96-106

K
Kadaba, M.P., 23, 60
Kinematic measurements, 16-17
    angular, 15, 32-35
        definition of anatomical joint angles, 32
        definition of segment Euler angles, 32-33
        Gaitmath mathematics for, 93-96
        velocity and acceleration of, 33-35
    linear, 15, 22-29
        for clinical subject, 64-65
        of clinical subject, 66-67
        determination of segment orientation, 27-29
        Gaitmath mathematics for, 87-91
        marker placement, 25-27, 87
        prediction of joint centers, 25-27
        use of markers, 23-24, 87
Kistler Force Plate (Kistler Instrument Corporation), 118-119
Kistler Instrument Corporation, 118-119
Knee. *See also* Calf
    axis of rotation for, 32, 33
    of clinical subject, 74, 75
    diameter of, 17-18
    joint forces and moments at, 38, 40
    pathway of, 1
    predicting position of joint center, 26, 88-89
    reference axes for, 41
    reference frame for, 32, 89-90
Knee diameter, 17-18
Knee extensor muscle, 13

L
Lambrinudi, C., 63
Lanshammar, H., 109-110
Lateral malleolus, 12-13, 24, 25, 26
Leg. *See* Calf; Knee; Thigh
Leonardo da Vinci, 78
Levine, W.S., 79
LIDO Kinetic Analysis System (Loredan), 116
Linear envelope, for processing EMG signals, 51-52, 65
Line of nodes, 33, 95-96
Loading matrix, 58
Loading response, 11, 54
Loading response factor, 59
Locomotion. *See* Gait (human); Motion; Movement
Loeb, G.E., 50, 51, 79, 108
Loredan, 116
Lurching, 13, 67, 69, 74

M
McConville, J.T., 19, 83-87
Macmillan, N. H., 78
Magnetic resonance imaging, 16-17
Malleolus height, 17-18
Malleolus width, 17-18
Man.ANG file (*GaitLab*), 6, 32, 34
Man.APM file (Gaitmath), 5, 17, 19, 65
Man.BSP file (Gaitmath), 5, 21-22
Man.COG file (*GaitLab*), 5-6, 30-31
Man.DYN file (*GaitLab*), 6, 41, 42
Man.FPL file (*GaitLab*), 5, 36, 37, 40, 65
Man.JNT file (*GaitLab*), 5, 27, 28
Mann, R.A., 55
Man.REF file (*GaitLab*), 5, 91
Marey, Etienne Jules, 78, 118
Mathematical modeling, of body segment parameters, 16-17
Measurement of human gait, 4-6, 15-43. *See also* Equipment for gait analysis
Metatarsals, 25
Midstance, 10-11, 54, 55, 59
Midswing, 11, 54, 55
Midthigh circumference, 17-18
MIE Medical Research Limited, 108, 110
Moments of inertia
    definition of, 16
    prediction of, 20-22
    relationship to ground reaction force, 80
    relationship to joint force, 40-43
Morlock, M., 20, 87
Motion, Newton's laws of, 37, 38, 39, 40, 96, 99-106

Motion Analysis Corporation, 23, 113-114, 124
Motion Analysis System Engine (Motiontronics for Science), 116-117
Motion artifact noise, 47, 51
Motiontronics for Science, 116-117
Motor action potential (MAP), 48
Motor unit, 48
Motor unit action potential (MUAP), 48
Movement
   as component in walking, 2-3
   direct dynamics of, 4
   inverse (indirect) dynamics of, 4-6, 15, 79
MOVIAS, Moving Image Analysis Software (Instrumentation Marketing Corporation), 112-113
MT8 Radio Telemetry System (MIE Medical Research Limited), 108
Multidimensional scaling, 57-58, 60-61
MULTILab (Selective Electronics, Inc.), 117
Muscles
   action versus contraction of, 49
   as component in walking, 2-3
   concentric action of, 49
   eccentric action of, 49
   electrophysiology of, 48-49
   interaction between, 55-61
   mapping of, 57-58
   measuring activity of, 13-14
   measuring tension in, 43
   and muscular dystrophy, 3
   phasic behavior of, 52-55
   and stability of joints, 80
   as synergists versus antagonists, 55
Muscular dystrophy, 3
Musculoskeletal system, as component in walking, 2-3. *See also* Muscles; Skeleton
Musgrave Footprint (WM Automation), 123
Muybridge, Eadweard, 78, 81

N
NAC Film Analyzer (Instrumentation Marketing Corporation), 112-113
National Institutes of Health (NIH) Biomechanics Laboratory, 24
Nerve impulses, 48
Nervous system, central
   and cerebral palsy, 3
   as component in walking, 2-3
   and control of muscle coordination, 14, 52
   and impulse for walking, 2
Nervous system, peripheral
   and Charcot-Marie-Tooth disease, 3
   as component in walking, 2-3
Neurectomy, 3
Newton, Sir Isaac, 37, 38, 39, 40, 96, 99-106
Norman, R.W., 30
Northern Digital, Inc., 117-118
Novel Electronics, Inc., 121
Numerical differentiation, 30, 93

O
OPTOTRAK system (Northern Digital, Inc.), 118
Orientation, of body segments, 27-29
Orthomat (Polysens SpA), 120-121
OrthoTrak (Motion Analysis Corporation), 23, 114, 124
Osteoarthritis, 10
OsteoKinetics Corporation, 125
Osteotomy, 4
Oxford Orthopaedic Engineering Centre (OOEC), 24

P
Pathologies of human gait, 3-4, 7, 10, 63-76. *See also names of individual medical conditions and diseases*
Patla, A.E., 60
Pattern recognition algorithms, 58
Paul, J., 55

Peak Performance Technologies, Inc., 113
Peak Video Motion Measurement Systems (Peak Performance Technologies, Inc.), 113
Pelvis. *See also* Hip
   markers for kinematic measurement of, 26-27, 89
   predicting position of hip joint center, 26-27
Penny & Giles Blackwood, Ltd., 111-112
Peroneus longus, 58, 59
Perry, 11
Perspiration, 47
Pezzack, J.C., 30
Plantar flexion, 32, 55, 71, 73-74, 95
Polysens SpA, 120-121
Pressure insoles, 118, 120-123
Preswing, 11, 54
Procter, P., 55
Propulsion factor, 59

Q
Quadriceps, 53
Quintic spline, 30, 31, 91-93

R
Radar, C.M., 91
Ralston, H.J., 1, 52, 53, 55, 78
Ramakrishnan, H.K., 23
Rancho Los Amigos Hospital, 11
Reaction boards, 16
Rectification, full-wave, 51-52
Rectus femoris, 13-14, 52, 56, 57, 59, 68, 70, 74
Recurvatum, 67, 74
Reynolds, H.M., 19, 83-87
Rheumatoid arthritis, 3
Rhizotomy, 3
Richmond Children's Hospital, 24
Right-handed screw rule, 25, 28-29
Rigid skeletal links
   assumption of rigidity for, 16
   as component in walking, 2-3
Roffey, M., 72
Rotation
   external, 12, 21, 22, 32, 33, 41, 93-95
   internal, 12, 21, 22, 32, 33, 41, 93-95

S
Sacrum, 26-27
Sagittal plane, 7, 25, 67, 79-80
Sartorius, 57, 59
SAS statistical package, 59
Scale, bathroom, 109
Science Accessories Corporation, 112
Second double support phase, 9-10
Selective Electronics, Inc., 117
Selspot II Cameras (Selective Electronics, Inc.), 117, 125
Shank. *See* Calf
Silver chloride, 47, 48, 51, 65
Single limb stance phase, 9-10
Skeleton, indirect measurement of, 25, 27
Software packages, integrated, 123-125
Soleus, 61
Spastic hemiplegia, 50
Speed walking, 81
Stance phase, 9-10, 22, 36-37, 80
Stanford, Leland, 78
Step width, 12
Stride length, 11-12
Suntay, W.J., 32, 93
Sutherland, D.H., 74
Sweat, 47
Swing phase, 9-10, 11, 80
Synge, J.L., 33, 96
Synovial joints
   as component in walking, 2-3
   and rheumatoid arthritis, 3

T
Tape measure, 17, 18, 64, 109
Tekscan, 122-123
Tenotomy, 3
Terminal stance, 11, 54, 59
Terminal swing, 11, 54
Theorem of Shannon, 49
Thigh
   center of gravity of, 30, 91
   joint dynamics of, 101-102, 104-105
   length of, 17-18
   orientation of, 28
   predicting mass of, 18-20
   predicting moments of inertia of, 21
Thigh length, 17-18
Threshold detector, 51-52
Tibialis anterior, 52, 53, 55, 57, 58, 59, 69-71
Tibialis posterior, 50
Tibial tubercle, 26
Todd, F., 1, 52, 53, 55, 78
Toe-off, 10-11, 54, 55, 60
Toes, clawing of, 63
Transkinetics telemetry system, 65, 108-109
Transverse plane, 7
Triax Three Dimensional Electro-Goniometer (Chattecx Corporation), 110-111
Triceps surae, 10, 53, 55, 57, 59, 69, 71
Turning point, 30-31
2K-QDA (Human Performance Technologies, Inc.), 115-116
2P-QDA (Human Performance Technologies, Inc.), 122

U
University of Florida at Gainesville, 124

V
Valgus, 95
van den Bogert, A.J., 79
Varus, 95
Vastus lateralis, 59
Vaughan, C.L., 4, 30, 72, 93
Velocity
   angular, 32, 33-35, 36, 96
   and calculating centers of gravity, 29-31, 93
VICON System (Oxford Metrics Ltd.), 24, 64-65, 114-115, 124-125
Videomex-X (Columbus Instruments), 116
Voyager Company, 78

W
Walking. *See* Gait (human)
Walking, speed, 81
Warner, A., 47
WATSMART (Northern Digital, Inc.), 117-118
Winter, D.A., 30, 41, 56-59, 61, 65, 67, 79, 91
WM Automation, 123
Woltring, H.J., 109
Wootten, M.E., 23, 60

Y
Yeadon, M.R., 20, 87
Young, J.W., 19, 83-87

Z
Zoetrope, 78